Chemical Kinetics

A Modern Survey of Gas Reactions

Chemical Kinetics

A Modern Survey of Gas Reactions

John Nicholas
University of London King's College

Harper & Row, Publishers
London New York Hagerstown San Francisco

First published 1976

Harper & Row Ltd
28 Tavistock Street, London WC2E 7PN

Designed by 'Millions'
Typeset by Cotswold Typesetting Ltd, Gloucester
Printed by R. J. Acford Ltd, Chichester, Sussex

Standard Book Number 06-318041-3

Contents

Chapter 5: Theories of Reaction Rates

Chapter 6: Radical Reactions, Non-Chain and Straight-Chain Reactions

Chapter 7: Branched-Chain Reactions

Chapter 8: Molecular Dynamics

Chapter 9: Molecular Beams

Chapter 10: Chemiluminescence, Hot-Atom Reactions, Ion–Molecule Reactions

Preface

Investigating rates of reactions and trying to understand such processes at the molecular level have formed an important part of chemistry for over a century. In the last few decades chemical knowledge has increased ever more rapidly, mechanisms of the most complex gas reactions have been established more firmly and new experimental and theoretical methods now probe homogeneous gas reactions in unprecedented microscopic detail. The teaching of kinetics should reflect both its more familiar basis (the 'conventional' kinetics which itself is resurgent at present) and also put into full and rich perspective the new generation of kinetic studies that has matured in the last few years. In addition, students would like some indication of how 'academic' kinetics contributes to the development of our industrial society and solves some of the associated environmental problems.

I hope this book in reflecting such aims will serve as a gas kinetics text for undergraduates throughout their degree course and provide for graduate students an introduction to any unfamiliar material. The text covers such traditional topics in experimental and theoretical kinetics as: the activation of molecules; conventional experimental methods; simple collision and activated complex theories; and complex reactions—linear and branched chain systems. I have also described modern techniques for investigating fast reactions, the rapidly expanding fields of molecular beam studies and molecular dynamic theories, and given a brief survey of ion–molecule reactions, hot atom reactions and chemiluminescence studies. Kinetic systems chosen for discussion include some which are important environmentally and in industry: hydrocarbon combustion and automobile engines; photochemical smog; ozone in the stratosphere; and ionic processes in the upper atmosphere.

I have tried to organize the textual material so that teachers will have great flexibility in selecting chapters or sections for their undergraduate courses. The only prerequisite for most of chapters 1–7 is an introductory physical chemistry course giving some basic knowledge of molecular structure and thermodynamics, together with some elementary calculus. Part of the theoretical treatment in chapter 5 invokes some statistical mechanics. The section on potential energy surface calculations and chapters 8–10 seem most suited to final year degree work, when students have greater background knowledge of molecular structure. The text may be followed partly or wholly in the order given, but many alternative arrangements are possible. For example, introductory descriptive kinetics in chapter 1 could be followed by molecular activation from chapter 2, conventional experimental methods (chapter 3), then simple collision and activated complex theories and unimolecular reactions from chapter 5, and complex reactions (chapters 6, 7)—alternatively chapters 6 and 7 could precede the rate theory in chapter 5. This leaves the sections in chapter 2 on excited states and energy transfer (a rapidly expanding field that is only briefly surveyed here) and chapter 4 on fast reactions for later study. A final course could embrace potential energy surface calculations from chapter 5, molecular beam studies and molecular dynamic theories (chapters 8, 9), and topics from chapter 10.

The problems, with answers, placed at the end of chapters are designed to re-inforce and sometimes extend the ideas presented there, and suggested further reading will introduce the reader to the wider range of kinetics literature. The book employs SI units, but students who have not been taught directly in this system should have no difficulty with the most common units: for concentration (mol cm^{-3}); energy (joules); atomic size and wavelength (nm). Teaching experience suggests that students do not change to SI pressure units with such easc so, though a clumsy device, both torr and the N m^{-2} equivalent are given.

King's College, London
June 1975

JN

Chapter 1: Basic Kinetic Laws

This chapter outlines the kinetic treatment appropriate for the more familiar type of gaseous system—the reactions of dilute gases (at pressures up to atmospheric) which, to a good approximation, are in thermal equilibrium at the temperature of the surroundings (usually between room temperature and 1 000 K). Such systems may be designated 'bulk' gases because at these pressures we are dealing with around 10^{19} molecules in every cubic centimetre and each molecule undergoes about 10^{10} collisions every second. Thus our kinetic description refers to the average behaviour of very large numbers of molecules in many molecular collisions and this provides a suitable basis for treating the gas reactions discussed in the first seven chapters. Recent developments in experimental and theoretical kinetics allow the investigation of reactants and products which are in specific energy states rather than at thermal equilibrium, systems where observed products result from single reactive collisions. For these it is essential to devise an alternative kinetic approach and such aspects of modern kinetics are discussed in the later chapters.

1.1 Rate of Reaction, Order of Reaction, Reaction Rate Constant

The overall course of a chemical reaction may be represented by a stoichiometric equation:

$$aA + bB + \ldots \rightarrow pP + qQ + \ldots \tag{1.1.1}$$

where a, b . . . and p, q . . . denote the number of moles of reactants (A, B . . .) and products (P, Q . . .).

Kinetics deals with the speed or rate at which reactions proceed and the *rate of reaction* is defined simply as the rate of change of concentration for reactants or products. So the reaction rate may be written in several ways which are related by the stoichiometry of the reaction. For example, the rate, R, of the above reaction may be expressed:

$$R = -\frac{d[A]}{dt} = -\frac{a}{b}\frac{d[B]}{dt} = \frac{a}{p}\frac{d[P]}{dt} = \frac{a}{q}\frac{d[Q]}{dt} \tag{1.1.2}$$

where the terms in square brackets are the concentrations of each species. The negative signs indicate that during reaction the concentrations of reactants decrease. The units of concentration may be number of moles per unit volume or number of molecules per unit volume, differing by Avogadro's number, $L = 6{\cdot}02 \times 10^{23}$. The basic unit of volume in SI units is m^3, but it is more conventional to use dm^3 ($1\ dm^3 \equiv 1$ litre $\equiv 10^{-3}\ m^3$) or $cm^3 \equiv 10^{-6}\ m^3$. Probably the most frequently encountered concentration units in gas kinetics are numbers of moles per cm^3 (mol cm^{-3}) which will usually be employed here. The corresponding units for rate are mol $cm^{-3}\ s^{-1}$.

The rate of a chemical reaction depends on several factors in addition to reactant concentration. In some cases the rate is influenced by the products, substances such as catalysts, or even chemically inert species. Of greater general importance is the fact that in almost all cases the rate varies with temperature, often very considerably. These important topics are discussed later, and the present section focuses on the dependence of rate on reactant concentration.

The variation of the experimentally measured rate with reactant concentration may often be expressed as in equation (1.1.3) though many reactions require more complicated equations

$$R = k[A]^x[B]^y \ldots \tag{1.1.3}$$

This is the *experimental rate equation* and it shows that the rate is usually proportional to each reactant concentration raised to some power. These experimentally determined exponents x, y . . . are called the *partial orders of reaction*: x with respect to A; y with respect to B. The sum of the partial orders, $x + y + \ldots$, is the *overall order of reaction*. The orders of reaction are often integers, but are not necessarily so. For the great majority of reactions these numbers are between 0 and 3. As will be seen later, experimental measurement of reaction order can provide important insights into the molecular detail of a reaction.

The proportionality constant, k, in the rate equation, is called the *reaction rate constant*. It is a fundamental kinetic parameter and the faster the reaction, the higher is its value. The rate constant varies with temperature and this is often emphasized by writing $k(T)$. A preliminary indication of the importance of this temperature variation is provided in section 10 of this chapter. Equation (1.1.3) shows that the units of k depend on the order of reaction. If the overall order $x + y + \ldots = Z$, then the units of k are $mol^{-(Z-1)}\ (cm^3)^{(Z-1)}\ s^{-1}$.

1.2 Complex Reactions, Reaction Mechanism

A reaction of simple stoichiometry such as:

$$A + B \rightarrow C + D \tag{1.2.I}$$

may be first order in A and first order in B. Further detailed investigation may reveal that the reaction does, in fact, proceed by a direct molecular interaction between just two species as indicated. Such a reaction may then be described as simple.

The rate equation for many reactions is more complicated than that described and the order of reaction may be fractional. Complicated rate equations may be found even where the stoichiometry of the overall reaction is simple. This indicates that the molecular detail of the reaction is more intricate than suggested by the overall reaction. In fact, this is the case for virtually all reactions in the gas phase, even those with simple orders of reaction. It can be demonstrated that, even where the overall kinetics are simple, reactants are transformed into products via a series of intermediate steps involving intermediate chemical species. Thus chemical reactions in the gas phase are usually *complex reactions*.

The series of intermediate steps which correspond to the overall reaction is often referred to as the *reaction mechanism*, for example, reaction (1.2.I) could proceed by the following mechanism:

$$A \rightarrow X + X \qquad (1.2.\text{II})$$

$$X + B \rightarrow Y + C \qquad (1.2.\text{III})$$

$$Y + A \rightarrow X + D \qquad (1.2.\text{IV})$$

$$X + X \rightarrow A \qquad (1.2.\text{V})$$

Initially A breaks up to give intermediate X which reacts with B to give product C and a second intermediate Y. Y then reacts with A to given product D and another X. The reaction sequence may be repeated until interrupted by the recombination of X intermediates to re-form A. Inspection of this mechanism reveals that it corresponds to the overall reaction, and it could result in a simple order of reaction (though for this particular example, see section 1.9). Thus kinetic measurements alone cannot establish the reaction mechanism unambiguously, independent evidence on the nature of the intermediates and the rates of intermediate reactions is required.

1.3 Elementary Reactions

Reactions (1.2.II–V) above are examples of intermediate reactions which comprise the mechanism of a complex reaction. These intermediate reactions have been shown by a wealth of experimental evidence to be 'simple', with simple orders corresponding to the reaction as written. The basic reactions are termed *elementary reactions*. They are either: first order, of stoichiometry $L \rightarrow$ products (e.g. reaction (1.2.II)); second order, $L + M \rightarrow$ products (e.g. reactions (1.2.III, IV)); or third order, $L + M + N \rightarrow$ products. There are, however, further complications in the first- and third-order cases which are discussed in chapter 5.

One of the major objectives of kinetics is to help deduce the sequence of elementary reactions that forms the mechanism of a complex reaction. The elementary reactions themselves are of great intrinsic importance, and another major task of the kineticist is to study the detailed molecular mechanics of these processes.

1.4 Molecularity

Molecularity may be defined as the number of species involved in the individual molecular process in which reactants are transformed into products. Strictly speaking the term can only be applied to elementary reactions: unimolecular, $L \rightarrow$ products; bimolecular $L + M \rightarrow$ products; termolecular, $L + M + N \rightarrow$ products. Hence the term is a truism and not very useful. It must not be confused with order of reaction which is an experimentally determined quantity. Further confusion may arise through specifying the molecularity of an overall complex reaction. Such use of the term can give no information on the reaction mechanism and can be thoroughly misleading. However, by historical convention, molecularity is still frequently employed in certain cases, in particular for reactions which display first-order kinetics and whose overall stoichiometry corresponds to a unimolecular equation.

1.5 Integrated Rate Equations

In kinetic experiments, the concentrations of the species present are measured at a series of times. These data may then be fitted to the appropriate experimental rate equation, such as:

$$R = -\mathrm{d}[A]/\mathrm{d}t = k[A]^x[B]^y$$

This rate equation is expressed in differential form, but it is often more convenient to use the corresponding integrated form. Such integrated rate equations are illustrated below for a few useful simple cases.

Zero-order reactions

Examples include the decomposition of some gases, such as ammonia, on metal catalysts. For the overall reaction: $A \rightarrow$ products, the differential rate equation is:

$$R = -\mathrm{d}[A]/\mathrm{d}t = k[A]^0$$

since the order of reaction is zero. $[A]^0$ is unity and by simple rearrangement

$$-\mathrm{d}[A] = k\,\mathrm{d}t.$$

Integrating gives:

$$-[A] = kt + \text{constant}$$

The integration constant is conveniently evaluated by taking the initial conditions $[A] = [A]_0$ at $t = 0$. Hence

$$[A] = -kt + [A]_0 \qquad (1.5.1)$$

So in the zero-order case a plot of concentration versus time gives a straight line graph of slope $(-k)$. The units of the zero-order constant are mol cm^{-3} s^{-1}.

First-order reactions

Complex gas reactions with first-order kinetics include the decomposition of nitrogen pentoxide, and there are many elementary reactions of this type. For the overall reaction: $A \rightarrow$ products, the differential rate equation is:

$$R = -\mathrm{d}[A]/\mathrm{d}t = k[A]$$

Integrating gives:

$$-\ln[A] = kt + \text{constant}$$

Evaluating the integration constant from initial conditions gives:

$$\ln[A] = -kt + \ln[A]_0 \tag{1.5.2}$$

So for first-order kinetics a plot of the logarithm of concentration versus time gives a straight line graph of slope $(-k)$. The units of the first-order constant are s^{-1}.

Equation (1.5.2) may be re-arranged to give:

$$[A] = [A]_0 \, e^{-kt} \tag{1.5.3}$$

which emphasizes that in first-order reactions the decay in reactant concentration is exponential.

Second-order reactions

(a) Type I

This involves the interaction of two molecules of the same reactant, as in the decomposition of hydrogen iodide and in many elementary reactions. For the overall reaction: $2A \rightarrow$ products, the differential rate equation is:

$$R = -\mathrm{d}[A]/\mathrm{d}t = 2k[A]^2$$

The factor 2 arises since every time the stoichiometric reaction takes place *two* molecules of A are removed. Integrating gives:

$$\frac{1}{[A]} = 2kt + \text{constant}$$

Evaluating the integration constant from initial conditions gives:

$$\frac{1}{[A]} - \frac{1}{[A]_0} = 2kt \tag{1.5.4}$$

Hence, for this type of second-order kinetics a plot of the inverse of concentration versus time gives a straight line graph of slope $2k$. The units of the second-order constant are $\text{mol}^{-1}\ \text{cm}^3\ \text{s}^{-1}$.

(b) Type II

This is first order in each of two reactants, for example the reaction between hydrogen and iodine and very many elementary reactions. For the overall reaction: $A + B \rightarrow$ products, the differential rate equation is:

$$-\mathrm{d}[A]/\mathrm{d}t = k[A][B]$$

If the initial concentrations of A and B are the same the treatment simplifies to that of type I. Where the initial concentrations differ, integration by parts and evaluation of the integration constant from initial conditions gives:

$$\frac{1}{[A]_0 - [B]_0} \ln\left(\frac{[A][B]_0}{[A]_0[B]}\right) = kt \tag{1.5.5}$$

Thus a graph of the logarithmic term versus time gives a straight line of slope $k([A]_0 - [B]_0)$.

Third-order reactions

The algebra involved in integrating the rate equation becomes more difficult and several different cases may be distinguished. The most important is encountered in elementary steps where two atoms combine in the presence of another molecule, and in the reaction of two nitric oxide molecules with one of oxygen.

For the overall reaction: $2A + B \rightarrow$ products, the differential rate equation is:

$$-\mathrm{d}[A]/\mathrm{d}t = 2k[A]^2[B]$$

which gives the integrated equation:

$$\frac{1}{[A]_0 - 2[B]_0}\left(\frac{1}{[A]_0} - \frac{1}{[A]}\right) + \frac{1}{([A]_0 - 2[B]_0)^2}\ln\left(\frac{[A][B]_0}{[A]_0[B]}\right) = kt \qquad (1.5.6)$$

the increase in complexity being very evident. The units of k are $\mathrm{mol^{-2}\ cm^6\ s^{-1}}$.

For other cases of integrated rate equations the reader is referred to standard compilations such as those cited at the end of the chapter.

1.6 Fractional Lifetimes

A useful indication of the relative speeds of reactions, under given experimental conditions, is the time taken for the reactant concentrations to fall to a specified fraction of their initial value. A popular choice is for reactants falling to half their initial concentration—the *half-life* for a reaction, $t_{\frac{1}{2}}$. Fractional lifetimes are related to the initial concentrations of reactants and the rate constant for the reaction. From the integrated rate equations considered above it is easily shown that: for zero-order reactions $t_{\frac{1}{2}} = [A]_0/2k$; for first-order, $t_{\frac{1}{2}} = \ln 2/k$; and for second-order type I, $t_{\frac{1}{2}} = 1/[A]_0 k$. For second-order type II and higher orders the corresponding expressions are more complicated. Inspection of the results quoted shows that for an nth-order reaction with single reactant, A, $t_{\frac{1}{2}} \propto 1/k[A]_0{}^{n-1}$.

For first-order reactions an additional point is that the half-life is independent of initial concentration. It was seen above that first-order decays are exponential (equation (1.5.3)) and another useful fractional lifetime is often quoted in such cases. This is the time, τ, for concentration to fall to $1/\mathrm{e}$ of the initial value. For first-order reactions

$$\tau = 1/k \qquad (1.6.1)$$

For a system at equilibrium which is disturbed and then returns to equilibrium by a first-order process τ is often referred to as a *relaxation time*.

1.7 Determination of Order of Reaction

It has already been stated that one of the primary aims of kinetic experiments is to find the order of reaction. Basic kinetic data consist of reactant concentrations at a series of times (at constant temperature). Using such data, orders of reaction may be determined by several methods.

(i) Use of integrated rate equations

In this method a reaction order is assumed, and a graph drawn for the corresponding integrated rate equation. This procedure is repeated until the order that gives the best fit is found. The method is not sufficiently precise to do more than, say, distinguish between first and second orders, and it is essential that the reaction is followed for several half-lives.

(ii) Use of fractional lifetimes

This method can be employed with reactions where the rate equation is of the form:

$$R = k[A]^n \tag{1.7.1}$$

In this case it has been shown already that the half-life is proportional to initial concentration: $t_{\frac{1}{2}} \propto 1/[A]_0^{n-1}$. This proportionality may be expressed in logarithmic form:

$$\log t_{\frac{1}{2}} = (1-n)\log[A]_0 + \text{constant}$$

Thus a plot of $\log t_{\frac{1}{2}}$ versus $\log [A]_0$ is a straight line whose slope gives n.

(iii) Differential method

Again where the rate equation is of the form (1.7.1), that equation may be expressed in logarithmic form:

$$\log R = n \log[A] + \log k$$

A plot of log (rate) against log (concentration) thus gives a straight line of slope n. The rates, however, must be measured from tangents to the concentration versus time curve, and the subjective nature of drawing tangents is a major disadvantage in this method.

(iv) Initial rates

This is the most generally useful method. Many reactions are complicated by processes involving reaction products and this is especially true in the gas phase. Such difficulties are minimized by the method that measures the rate at the start of the reaction and the reactant concentrations are known most accurately at this time. However, the disadvantage of measuring rates by drawing tangents also applies here.

Investigating the variation of initial rate with initial reactant concentrations allows the determination of partial orders. If the rate equation for the initial conditions is:

$$R_0 = k[A]_0^x[B]_0^y$$

then taking logarithms:

$$\log R_0 = \log k + x \log[A]_0 + y \log[B]_0$$

In a series of experiments at constant $[A]_0$, the variation of R_0 with $[B]_0$ may be established. A plot of $\log R_0$ versus $\log[B]_0$ thus gives partial order y. Conversely, partial order x may be measured at constant $[B]_0$. Ideally, for each initial rate, several concentration measurements should be made within the first few per cent reaction. If the reaction is too fast to allow this other methods may be more suitable.

The order measured at the initial time is sometimes called the true order or 'order with respect to concentration'. In other methods described, measurement extends over a large fraction of the reaction time.

(v) Method of isolation

Another method which yields partial orders of reaction can be used if it is possible to add all reactants but one in large excess. The concentration of the excess reactants remains approximately constant while the variation of the remaining reactant with time is studied. For example, if the experimental rate equation is:

$$R = k[A]^x[B]^y[C]^z$$

and A and B are added in large excess, then

$$R \sim k'[C]^z$$

The partial order z may then be obtained by one of the methods described above. The other reactants may similarly be studied 'in isolation'.

Measurement of rate constants

Rate constants are fundamental kinetic parameters for bulk systems and their evaluation is the special concern of the kineticist. All the methods for determining reaction orders discussed above also allow the measurement of rate constants.

If experiments of this type are repeated at other temperatures, the variation of the rate constant with temperature can be deduced. The importance of this determination is discussed in section 1.10 and in subsequent chapters.

1.8 Complex Reactions (Opposing, Concurrent, Consecutive)

Frequently reactions do not exhibit simple orders of reaction and many carefully planned experiments are necessary to find the appropriate rate equations. These rate equations provide essential evidence for deducing the mechanism of the complex reaction. The mechanisms encountered in gas kinetics are of almost infinite variety, but a few fundamental types may be postulated and discussing the kinetic features of these simple examples can contribute to understanding the considerable problems presented by complex reactions.

A convenient basic classification of complex reactions is where the participating elementary reactions are: (i) opposing (reversible reactions); (ii) concurrent; or (iii) consecutive. In practice complex reactions involve an intermixture of two or all three types. For simple examples of each type it is possible to write down differential rate equations and obtain the integral rate equations for reactants, intermediates and products. With more complicated cases it becomes increasingly difficult and often impossible to obtain exact integrated rate equations. Approximate methods have been developed for such reactions and they are discussed later.

(i) Opposing reactions

The major kinetic features are conveniently illustrated by the simplest example—opposing first-order reactions:

$$A \underset{k_2}{\overset{k_1}{\rightleftharpoons}} B$$

If initially only A is present, at concentration $[A]_0$, and an amount x has reacted at time t, then the differential equation:

$$\mathrm{d}[B]/\mathrm{d}t = k_1[A] - k_2[B]$$

may be re-written with $[A] = [A]_0 - x$ and $[B] = x$:

$$\mathrm{d}x/\mathrm{d}t = k_1([A]_0 - x) - k_2 x \tag{1.8.1}$$

Eventually the system will reach equilibrium, where the value of x is x_e. At equilibrium the forward and back reactions proceed at equal rates and so $\mathrm{d}x/\mathrm{d}t = 0$.

This is in accord with *the principle of microscopic reversibility* which was formulated by Tolman[1]. In essence it may be expressed: if in the final state after molecules collide all the molecular momenta, internal and translational, are reversed the system will return by the same path in the reverse direction to give the initial state with momenta reversed. This can be extended from the microscopic scale to large molecular assemblies at equilibrium to give *the principle of detailed balancing*: in a system at equilibrium any molecular process and its reverse proceed, on average, at the same rate.

Under these conditions

$$k_1([A_0] - x_e) - k_2 x_e = 0 \tag{1.8.2}$$

This relationship makes it possible to transform the differential rate equation into a form suitable for integration. Combining (1.8.1) and (1.8.2) gives

$$\frac{\mathrm{d}x}{\mathrm{d}t} = \frac{k_1[A]_0}{x_e}(x_e - x)$$

This is easily integrated, and the integration constant evaluated from the initial conditions $x = 0$, $t = 0$, giving:

$$\ln\left(\frac{x_e}{x_e - x}\right) = k_1 \frac{[A]_0}{x_e} t \tag{1.8.3}$$

Thus, from measurements of the concentration of A or B until equilibrium is reached, a graph of $\ln\{x_e/(x_e - x)\}$ versus t may be plotted giving a straight line of slope $(k_1[A]_0/x_e)$ and hence k_1. With this value for k_1, k_2 may be deduced from equation (1.8.2).

By definition the equilibrium constant, K_c, is given:

$$K_c = \frac{[B]_e}{[A]_e} = \frac{x_e}{[A]_0 - x_e} \tag{1.8.4}$$

From (1.8.2) and (1.8.4) ·

$$K_c = k_1/k_2$$

Thus the equilibrium constant is equal to the ratio of rate constants for the forward and back reactions. Such a relationship is general for opposing reactions in an equilibrium system.

[1] R. C. Tolman, *Phys. Rev.*, **23**, 699 (1924); *The Principles of Statistical Mechanics*, Clarendon Press, Oxford, 1938, p. 163.

Small displacement from equilibrium

An alternative way of studying opposing reactions is to start with the system at equilibrium, cause a small displacement, and observe the subsequent relaxation back to equilibrium.

Again taking the case of opposing first-order reactions, let the new equilibrium concentrations be $[A]_e$ and $[B]_e$. If the small displacement in concentration is y, then the rates of reaction are given by the differential equation:

$$\begin{aligned} dy/dt &= k_1([A]_e - y) - k_2([B]_e + y) \\ &= k_1[A]_e - k_2[B]_e - (k_1 + k_2)y \end{aligned}$$

The rates of opposing reactions at equilibrium are equal, so

$$k_1[A]_e - k_2[B]_e = 0$$

giving

$$dy/dt = -(k_1 + k_2)y \qquad (1.8.5)$$

Integrating (1.8.5), and evaluating the integration constant by taking $y = y_0$ (the initial displacement) at $t = 0$, yields:

$$\ln(y/y_0) = -(k_1 + k_2)t$$

or:

$$y = y_0 \, e^{-(k_1+k_2)t} \qquad (1.8.6)$$

Comparison of equation (1.8.6) with (1.5.3) emphasizes that the return of the system to equilibrium, or its *relaxation* to equilibrium, is by an exponential decay. The decay exhibits first-order kinetics with rate constant $(k_1 + k_2)$. The relaxation may also be characterized by the appropriate fractional lifetime for the process. In section 1.6 it was pointed out that the time required for the change to $1/e$ of the initial value was a useful parameter. Here $\tau = 1/(k_1 + k_2)$ and τ is called the *relaxation time.* So measurement of relaxation times in such systems gives $(k_1 + k_2)$ directly, and if the equilibrium constant is known both k_1 and k_2 may be calculated separately. Similar results are obtained for opposing second-order processes where the displacement from equilibrium is *small.*

(ii) Concurrent reactions

The main kinetic features of a system in which a reactant undergoes more than one process in parallel, may be illustrated for first-order reactions:

$$A \xrightarrow{k_1} B$$
$$A \xrightarrow{k_2} C$$

The differential rate equations for reactant and products are:

$$-d[A]/dt = k_1[A] + k_2[A] = (k_1 + k_2)[A] \qquad (1.8.7)$$

$$d[B]/dt = k_1[A] \qquad (1.8.8)$$

$$d[C]/dt = k_2[A] \qquad (1.8.9)$$

If initially only A is present at concentration $[A]_0$ equation (1.8.7) may be integrated to give the simple first-order equation

$$\ln([A]/[A]_0) = -(k_1 + k_2)t$$

or

$$[A] = [A]_0 \, e^{-(k_1+k_2)t} \tag{1.8.10}$$

Substituting into (1.8.8) gives, for B:

$$d[B]/dt = k_1[A]_0 \, e^{-(k_1+k_2)t}$$

which may be integrated to give

$$[B] = \frac{k_1}{k_1 + k_2}[A]_0(1 - e^{-(k_1+k_2)t}) \tag{1.8.11}$$

Similarly from (1.8.9)

$$[C] = \frac{k_2}{k_1 + k_2}[A]_0(1 - e^{-(k_1+k_2)t}) \tag{1.8.12}$$

The variation in concentration of A, B and C with time is given by plotting equations (1.8.10, 11, 12) and is illustrated in figure 1.1 for arbitrarily chosen values of k_1 and k_2. Inspection of the mechanism and kinetics shows that at all times $[B]/[C] = k_1/k_2$.

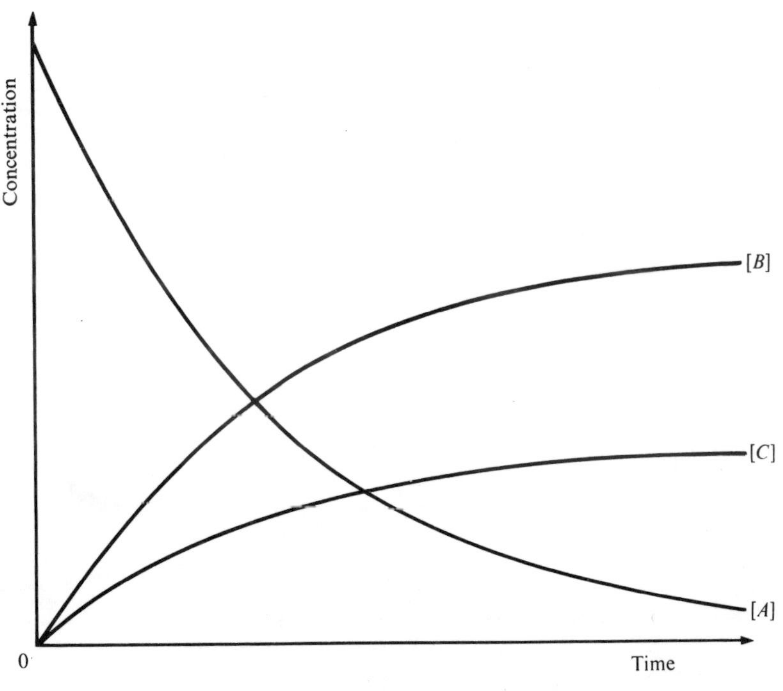

Figure 1.1
Variation with time of concentration of reactant, A, and products B, C in a system of two first-order concurrent reactions. $k_1/k_2 = 2$ in this example.

(iii) Consecutive reactions

This class of reactions is of great importance in gas kinetics, and again some salient features emerge from the simplest first-order case. The mechanism is:

$$A \xrightarrow{k_1} B$$
$$B \xrightarrow{k_2} C$$

and reactant A is converted to product C via intermediate B.

The differential rate equations are:

$$\mathrm{d}[A]/\mathrm{d}t = -k_1[A] \tag{1.8.13}$$

$$\mathrm{d}[B]/\mathrm{d}t = k_1[A] - k_2[B] \tag{1.8.14}$$

$$\mathrm{d}[C]/\mathrm{d}t = k_2[B] \tag{1.8.15}$$

If initially only A is present at concentration $[A]_0$ then (1.8.13) may be integrated to give the simple first-order equation:

$$[A] = [A]_0\, \mathrm{e}^{-k_1 t} \tag{1.8.16}$$

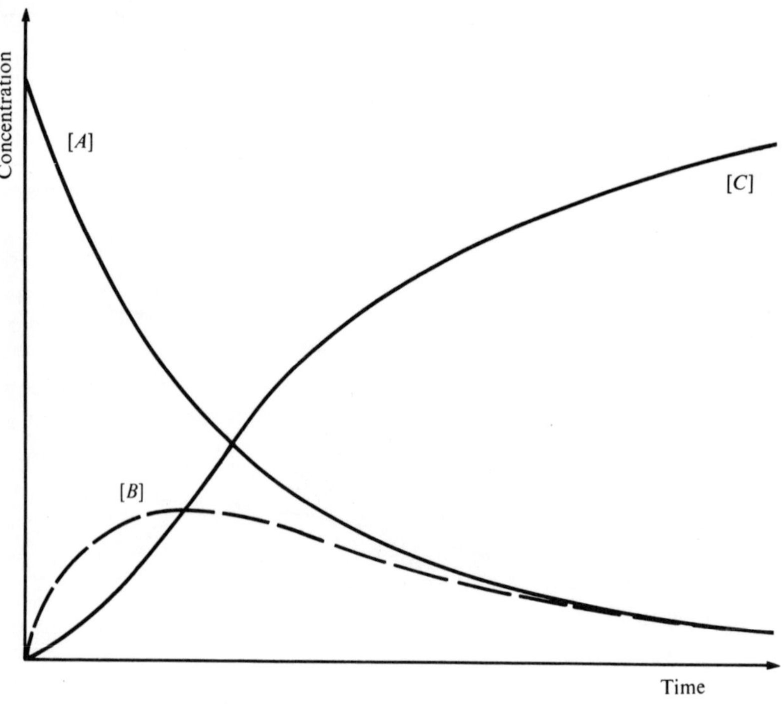

Figure 1.2
Variation with time of concentration of reactant A, intermediate B and product C, in a system of two consecutive first-order reactions where $k_2 = 2k_1$.

The differential equation for B becomes:

$$\mathrm{d}[B]/\mathrm{d}t = k_1[A]_0\,\mathrm{e}^{-k_1 t} - k_2[B]$$

and it may be integrated to give (1.8.17) below. This result can be checked most easily by differentiating (1.8.17)

$$[B] = \frac{k_1}{k_2 - k_1}[A]_0\,(\mathrm{e}^{-k_1 t} - \mathrm{e}^{-k_2 t}) \tag{1.8.17}$$

Only A was present initially, so $[A] + [B] + [C] = [A]_0$. Hence

$$[C] = [A]_0\left(1 - \frac{k_2\mathrm{e}^{-k_1 t} + k_1\mathrm{e}^{-k_2 t}}{k_2 - k_1}\right) \tag{1.8.18}$$

The variation of A, B and C with time is given by plotting equations (1.8.16, 17, 18) and is illustrated in figure 1.2 for the case where $2k_1 = k_2$.

Many kinetic systems involve highly reactive intermediates such as atoms or radicals. Within the terms of the present example this would correspond to intermediate B being formed relatively slowly from a stable molecule A, and then reacting very rapidly

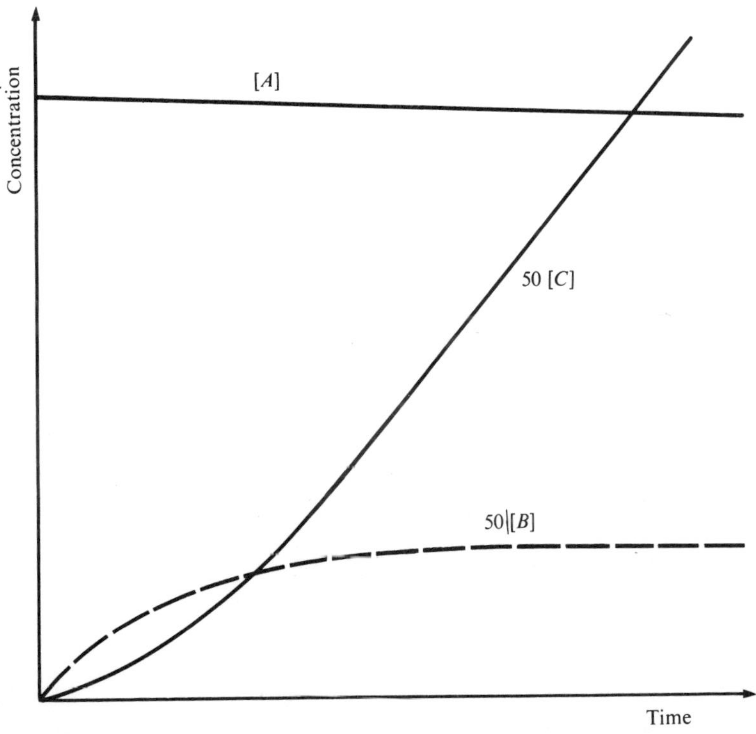

Figure 1.3
Variation with time of concentrations of A, B, C in a system of two consecutive first-order reactions where $k_2 = 200\,k_1$.

to give product C. Thus $k_2 \gg k_1$. With more complex mechanisms it remains generally true that reactive intermediates are formed slowly and then react far more rapidly. The consequences for the present example are illustrated in figure 1.3 for the case $k_2 = 200\ k_1$. The scale of B and C have been magnified by $\times 50$ for clarity. This figure indicates some points which are of general importance.

(i) The concentration of the reactive intermediate soon reaches an almost steady value, but this concentration is, at all times, very low.

(ii) The greater the relative magnitude of k_2, the quicker this approximately *steady state* is reached.

(iii) The increase in concentration of product C soon becomes almost linear. Thus the rate of formation of C ($\mathrm{d}[C]/\mathrm{d}t$, the slope of this curve) becomes approximately *steady*.

In this simplest case of consecutive reactions it is possible to derive exact integrated rate equations. Most real examples of complex reactions with consecutive elementary steps are far more complicated and finding exact integrated equations becomes prohibitively difficult. Approximate methods for finding suitable rate equations must be employed and that most widely used is described below.

1.9 The Steady-State Approximation

In most complex gas reactions the intermediates are highly reactive radicals. Their rate of formation is relatively slow but they react very rapidly. On the pattern indicated above they soon attain an approximately steady concentration that is very low compared with the concentration of stable reactant and product molecules, too low for convenient measurement. Similarly the rate of product formation soon reaches an approximately steady value. It is usually convenient to measure experimentally the rate of reaction at its approximately steady value and to see how this *steady-state* rate varies with the concentration of stable reactants.

For comparison with the experimental steady-state rate equation, it is possible to derive a rate equation from the mechanism of the complex reaction. This derivation is made feasible by use of the *steady-state hypothesis* (or *stationary-state hypothesis*). This simply states that during the steady-state period the concentration of reactive intermediates is approximately constant and their rate of change of concentration is approximately zero.

In the first-order case above this corresponds to:

$$[B]_{ss} = \text{constant} \quad \text{and} \quad (\mathrm{d}[B]/\mathrm{d}t)_{ss} = 0$$

The steady-state rate equation in this simple example is expressed as the rate of formation of products during the steady period $(\mathrm{d}[C]/\mathrm{d}t)_{ss}$. By putting equation (1.8.14) equal to zero this gives:

$$(\mathrm{d}[C]/\mathrm{d}t)_{ss} = k_2[B]_{ss} = k_1[A]$$

It is instructive to compare this result with the exact solution, using equation (1.8.17):

$$\frac{d[C]}{dt} = k_2[B] = \frac{k_2 k_1}{k_2 - k_1}[A]_0(e^{-k_1 t} - e^{-k_2 t})$$

For the steady state to be a reasonable approximation, $k_2 \gg k_1$. Applying this condition to the exact equation above, and incorporating equation (1.8.16) gives $(d[C]/dt)_{ss} = k_1[A]$, the same as the result obtained directly from the steady-state hypothesis.

As an example of a somewhat more realistic mechanism, though still far simpler than found in most important gas reactions such as those discussed in chapters 6, 7, consider the overall reaction:

$$A + B \rightarrow C + D$$

which proceeds by the mechanism

$$A \xrightarrow{k_1} X + X$$
$$X + B \xrightarrow{k_2} Y + C$$
$$Y + A \xrightarrow{k_3} X + D$$
$$X + X \xrightarrow{k_4} A$$

X, Y are reactive intermediates to which the steady-state hypothesis is applicable:

$$d[X]/dt = 2k_1[A] - k_2[X][B] + k_3[Y][A] - 2k_4[X]^2$$

$$d[Y]/dt = k_2[X][B] - k_3[Y][A]$$

Setting these equations equal to zero gives by simple algebra:

$$[X]_{ss} = \left(\frac{k_1[A]}{k_4}\right)^{1/2} \quad \text{and} \quad [Y]_{ss} = \frac{k_2[B]}{k_3[A]}[X]_{ss}$$

The rate of formation of products is given, for example, by

$$d[C]/dt = k_2[X][B]$$

and under steady-state conditions

$$\left(\frac{d[C]}{dt}\right)_{ss} = k_2\left(\frac{k_1}{k_4}\right)^{1/2}[A]^{1/2}[B]$$

It is easily shown that in this case the same result is obtained for $d[D]/dt$, $-d[A]/dt$ and $-d[B]/dt$. This steady-state rate equation should be compared with the experimental rate equation to test the postulated mechanism. If the theoretical and experimental equations have the same form this supports the mechanism but is *not* sufficient to confirm it. Typical further supporting evidence would be to find independent values for the rate constants of the elementary reactions, k_1, k_2, k_4, for comparison with the overall rate constant $[k_2(k_1/k_4)^{1/2}]$.

Finally it must be emphasized that for any kinetic system the validity of the steady-state assumption should be tested as rigorously as allowed by knowledge of the rate constants of the elementary reactions. The method should only be applied if the steady state is set up in a time corresponding to the consumption of a very small fraction of reactants.

Non-steady-state conditions

Methods are available for interpreting complex kinetics based on rather less drastic assumptions, though they are necessarily more complicated to apply. Their main advantage would be their use in non-steady periods of the reaction. An example is given in section 7.2, and a fuller discussion can be found in standard texts cited at the end of chapter 7. There is also a progressive increase in the use of computers to obtain numerical solutions for complex systems.

1.10 Dependence of the Rate Constant on Temperature

It has been mentioned several times that kinetic experiments should be carried out at constant temperature since most rate constants vary greatly with temperature. This variation excited the interest of early kineticists who were able to establish an empirical relationship of the form $k = A\,\mathrm{e}^{-B/T}$ where A and B are constants[2]. Arrhenius and van't Hoff provided a vital insight into the energetics of reactions with the suggestion that normal molecules in the system existed in equilibrium with activated molecules, and it was the activated molecules which reacted[3]. This idea was incorporated into the empirical temperature equation by stating $B = E_{\mathrm{exp}}/R$, where R is the gas constant and E_{exp} is the amount of energy required to form the activated species from reactants. This results in the famous Arrhenius equation:

$$k = A\,\mathrm{e}^{-E_{\mathrm{exp}}/RT} \tag{1.10.1}$$

Thus the temperature dependence of the rate constant is exponential, and governed by the value of E_{exp}. This most important quantity is called the *Arrhenius activation energy*, or, perhaps more relevantly, the *experimental activation energy*. For most reactions its value is in the range 0–400 kJ mol^{-1}. In general the value of the exponential term, and so the rate constant, increases rapidly with temperature—a typical activation energy of about 50 kJ mol^{-1} means that in the region of room temperature the rate constant doubles in value when the temperature is raised by 10 K. A minor point is that the form of the Arrhenius equation shows that at very high temperatures the value of k tends to A as a limit. However, only for reactions of very low activation energy is this limit approached at experimentally accessible temperatures.

The constant A in the Arrhenius equation is usually called the *pre-exponential factor* or *frequency factor*. Since the exponential term is a number, the units of A are the same as the units of the rate constant: s^{-1} for the first-order constants; mol^{-1} cm^3 s^{-1} for second-order, etc.

The Arrhenius equation embraces all temperature variation within the exponential term. Some authors allow for a slight temperature variation (say $T^{1/2}$) in the pre-exponential factor, and in extreme cases in the activation energy term itself. It is very difficult to detect experimentally any temperature variation in the pre-exponential factor or the activation energy. For the present, the experimental activation energy will be defined by the Arrhenius equation (1.10.1). It may also be expressed in an

[2] J. J. Hood, *Phil. Mag.*, **6**, 371 (1878); *Phil. Mag.*, **20**, 323 (1885).

[3] J. H. van't Hoff, *Etudes de Dynamique Chimique*, Muller, 1884. S. Arrhenius, *Z. Phys. Chem.*, **4**, 226 (1889).

alternative form of this equation, which by simple algebraic re-arrangement gives:

$$E_{exp} = RT^2 \frac{d(\ln k)}{dT} \tag{1.10.2}$$

Pre-exponential factors and activation energies are obtained experimentally from measurements of the rate constant at a series of temperatures. The Arrhenius equation in logarithmic form is:

$$\ln k = \ln A - E_{exp}/RT$$

Hence a plot of $\ln k$ versus $1/T$ gives a straight line with slope $-E_{exp}/R$ and intercept $\ln A$, and this is illustrated in figure 1.4.

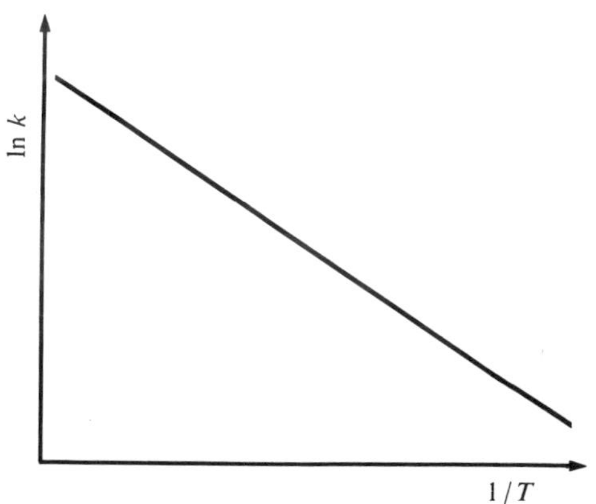

Figure 1.4
Variation of rate constant with temperature. An Arrhenius plot of $\ln k$ versus $1/T$.

This type of temperature variation governs reaction rates not only in the gas phase but also in solution and indeed in biological systems. Some amusing examples have been discussed by Laidler[4] including the observation that the flashing of fireflies obeys the Arrhenius law with an activation energy of about 50 kJ mol^{-1}. This information is not trivial since activation energies of this magnitude are characteristic of chemical processes, and it is probable that such phenomena are essentially chemical in nature.

Other aspects of van't Hoff and Arrhenius' arguments emerge from consideration of an equilibrium system:

$$A + B \underset{k_2}{\overset{k_1}{\rightleftharpoons}} C + D$$

for which:

$$K_c = k_1/k_2 \tag{1.10.4}$$

[4] K. J. Laidler, *J. Chem. Ed.*, **49**, 343 (1972).

The temperature variation of an equilibrium constant is given by the standard thermodynamic relationship (due to van't Hoff) in terms of ΔE, the difference in energy between reactants and products:

$$RT^2 \frac{d(\ln K_c)}{dT} = \Delta E \tag{1.10.5}$$

From (1.10.4) and (1.10.5)

$$RT^2 \frac{d(\ln k_1)}{dT} - RT^2 \frac{d(\ln k_2)}{dT} = \Delta E \tag{1.10.6}$$

But from (1.10.2) it can be seen that the left-hand side of (1.10.6) is $E_1 - E_2$ where E_1 and E_2 are the experimental activation energies for reactions 1 and 2. Hence:

$$E_1 - E_2 = \Delta E$$

This result is expressed on the energy diagram in figure 1.5. The diagram shows that the activated states for forward and back reaction are at the same energy, in fact for both forward and reverse reaction the system must pass through the *same* state of higher energy.

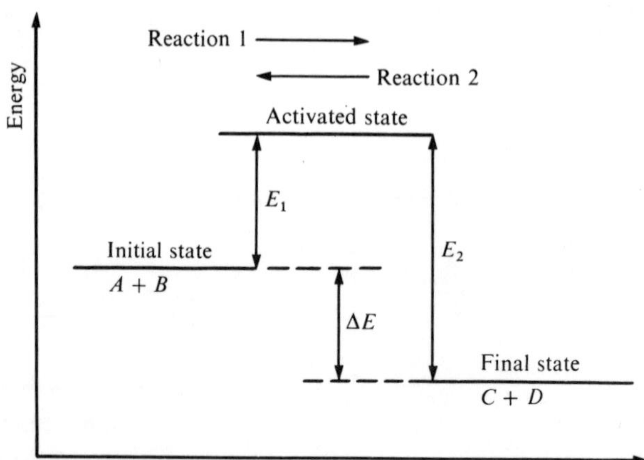

Figure 1.5
Schematic representation of changes in energy between initial, activated and final states of system $A + B \underset{2}{\overset{1}{\rightleftharpoons}} C + D$.

The study of thermodynamics shows that for a given initial state of the system, the final state eventually reached is determined by thermodynamic criteria, particularly the difference in energy ΔE. Thermodynamics gives no information on the *rate* at which the final state is reached. The rate of the processes involved is determined by kinetic criteria, particularly the activation energy. Furthermore the rate of the forward process depends on E_1 alone, not E_2 or ΔE, and the rate of the reverse reaction depends on its activation energy, E_2 alone.

The early part of this chapter introduced the role of the kineticist in deducing the mechanism of reactions and measuring their rate constants. The present section discusses the temperature variation of the rate constant in terms of the pre-exponential factor and activation energy. This suggests another type of problem for the kineticist—investigating the molecular detail of elementary reactions in order to understand the form of the Arrhenius equation and to predict the values of the fundamental kinetic parameters A, E_{exp} and so k. The theoretical approach to kinetics is described in chapters 5 and 8.

Further Reading

S. W. Benson, *Foundations of Chemical Kinetics*, McGraw-Hill, 1960.

C. Capellos and B. Bielski, *Kinetic Systems*, Wiley Interscience, 1972.

E. S. Swinbourne, *Analysis of Kinetic Data*, Nelson, 1971.

Comprehensive Chemical Kinetics, Vol. 1 (Eds C. H. Bamford and C. F. H. Tipper), Elsevier, 1969.

Techniques of Chemistry, Vol. 6, Part 1, 3rd ed. (Ed. E. S. Lewis), Wiley Interscience, 1974.

J. R. Hulett, *Quart. Rev.*, **18**, 227 (1964).

Exercises

In all examples reactions are studied at constant volume unless otherwise stated.

1.1
The first-order decomposition of a pure substance proceeds as shown in the table. Calculate the value of the rate constant.

Concentration/10^{-5} mol cm^{-3}	20·8	15·5	10·6	7·1	5·4	4·1	2·8	2·0
Time/s	0	50	108	146	188	246	298	350

(Ans: $6{\cdot}6 \times 10^{-3}$ s^{-1})

1.2
Investigate whether the reaction half-life is independent of 'initial' concentration for the reaction in exercise 1.1.

1.3
The reaction whose stoichiometry is $A + B \rightarrow$ Products is first order in A and first order in B. The initial concentration of B was $1{\cdot}01 \times 10^{-4}$ mol cm^{-3} and the concentration of A varied with time as shown in the table. Calculate the value of the rate constant.

Concentration/10^{-4} mol cm^{-3}	1·52	1·41	1·31	1·16	1·06	0·97	0·90	0·85	0·81
Time/s	0	300	600	1200	1800	2400	3000	3600	4200

(Ans: 2·7 mol^{-1} cm^{3} s^{-1})

1.4
In the second-order reaction with stoichiometry $2A \rightarrow A_2$ the total pressure varied with time as shown in the table. The temperature was 400 K. Evaluate the rate constant in pressure and concentration units.

Pressure/Torr	632	584	547	509	475	453	433	405	381
Time/s	0	24	49	83	136	180	238	353	519

(Ans: $1{\cdot}2 \times 10^{-5}$ Torr^{-1} s^{-1} = $2{\cdot}7 \times 10^{2}$ mol^{-1} cm^{3} s^{-1})

1.5
In the reaction between A and B the variation of initial rate with initial concentration of A, at constant initial concentration of B (10^{-5} mol cm^{-3}), is given in the following table:

Initial concentration of $A/10^{-6}$ mol cm^{-3}	10	7	4	2	1
Initial rate/10^{-8} mol cm^{-3} s^{-1}	31·6	22·1	12·7	6·32	3·16

When the initial concentration of A is kept constant at 10^{-5} mol cm^{-3} the variation of initial rate with initial concentration of B is given by:

Initial concentration of $B/10^{-6}$ mol cm^{-3}	10	7	4	2	1
Initial rate/10^{-7} mol cm^{-3} s^{-1}	3·16	2·65	2·05	1·41	1·0

What are the partial orders of reaction with respect to A and B and the value of the rate constant?

(Ans: Orders of reaction are 1 for A, 0·5 for B; $k = 10$ mol$^{-1/2}$ cm$^{3/2}$ s^{-1})

1.6
When a stoichiometric mixture of A and B react the concentration of A varies with time as shown in the table:

Concentration of $A/10^{-6}$ mol cm^{-3}	10·0	9·14	7·91	7·07	5·77	4·47	3·54	3·02	2·18
Time/s	0	10	30	50	100	200	350	500	1000

Use the differential method graphically to find the total order of reaction and the value of the rate constant. In an experiment in which A was present in large excess the half-life was independent of the initial concentration of B. What are the partial orders of reaction?

(Ans: Orders of reaction are 2 for A and 1 for B; $k = 10^{8}$ mol^{-2} cm^{6} s^{-1})

1.7
In the reaction

$$A + B \underset{k_2}{\overset{k_1}{\rightleftharpoons}} C$$

the equilibrium is subjected to a *small* disturbance and the system relaxes to a new equilibrium position where the concentrations of A and B are $[A]_e$, $[B]_e$. Show that the relaxation obeys first-order kinetics and that the relaxation time τ is given by $\tau^{-1} = k_1([A]_e + [B]_e) + k_2$.

1.8

For the system of consecutive unimolecular reactions, $A \xrightarrow{k_1} B \xrightarrow{k_2} C$, the integrated rate equation for $[B]$ is given by equation (1.8.17). Show that the maximum concentration of B occurs at $t_{max} = \ln\{(k_2/k_1)\}/(k_2 - k_1)$ and that $[B]_{max} = [A]_0(k_1/k_2)^{k_2/(k_2-k_1)}$. N.B. $e^{\ln x} = x$.

If $k_1 = 10^{-8}\ s^{-1}$ and $k_2 = 1\ s^{-1}$ evaluate t_{max}. From equation (1.8.17), remembering that $k_2 \gg k_1$, estimate the time, t', required for the concentration of B to reach $0{\cdot}9[B]_{max}$. What are the implications of these results for the steady-state hypothesis? (Ans: $t_{max} = 18{\cdot}5$ s; $t' = 2{\cdot}3$ s)

1.9

The overall reaction, $A + B \rightarrow C + D$ proceeds by the following mechanism: $A \xrightarrow{k_1} X + C$; $X + C \xrightarrow{k_2} A$; $X + B \xrightarrow{k_3} C + D$. The relative rates of these reactions are such that the steady-state hypothesis may be applied to intermediate X. Show that:

(a) If $k_2[C] \gg k_3[B]$ then $-d[B]/dt = k_1k_3[A][B]/k_2[C]$;

(b) If $k_2[C] \ll k_3[B]$ then $-d[B]/dt = k_1[A]$.

1.10

The temperature variation of the rate constant for an unimolecular reaction is given in the table. Evaluate graphically the experimental activation energy and pre-exponential factor.

T/K	500	520	540	560	580	600
$k/10^{-8}\ s^{-1}$	1·29	8·20	45·3	231	977	3 910

(Ans: $A_{exp} = 1{\cdot}0 \times 10^{13}\ s^{-1}$; $E_{exp} = 200$ kJ mol^{-1})

1.11

A reaction proceeds by the mechanism given in exercise 1.9, under conditions where $k_2[C] \gg k_3[B]$. Show that the overall experimental activation energy is related to the activation energies of the elementary reactions by $E_{exp} = E_1 + E_3 - E_2$.

E_{exp} is found to be 300 kJ mol^{-1}. Reaction 1 is 250 kJ mol^{-1} endothermic. What is E_3?

(Ans: 50 kJ mol^{-1})

Chapter 2: Molecular Activation

Gas phase reactions are complex, with a mechanism of elementary reactions which usually involve highly reactive radicals. A *radical* can be defined quite simply as a species with an unpaired electron. This definition includes polyatomic species such as $CH_3^{\cdot}$ and $ClO^{\cdot}$ and many atoms such as $H^{\cdot}$, or halogen atoms. The only anomaly in this definition is that a few molecules, notably NO and O_2, also contain unpaired electrons. Although elementary radical reactions are fast, the overall conversion of reactants to products may be comparatively slow. Such overall rates have been studied by kineticists since the beginning of this century, but in the last thirty years techniques have been developed for directly observing the fast radical reactions themselves.

Whether a kinetic experiment aims at measuring overall rates or rates for specific elementary radical reactions the method must provide a source of radicals. To form radicals, bonds in stable molecules must be broken, and since the energy required to break such bonds is usually in the range 200–500 kJ mol^{-1}, some energy source is essential to activate the molecules that dissociate. The molecular details of these processes leading to bond fission are fundamentally important in chemistry and they are discussed below with some orientation toward the special interests of kinetics.

The most common energy sources are thermal and photochemical, but radiochemical methods and electrical discharges are also used to some extent.

2.1 Radiochemical Activation

X-rays, high energy electrons and α particles are common types of high energy radiation. Their effect on molecules is drastic, giving electrons, ions, highly excited molecules and radicals. The energies involved are far higher than normally encountered in kinetics and so radiation chemistry is very much a separate subject which will not be considered further here.

2.2 Electrical Discharge Activation

When a high voltage discharges through gases at low pressure the primary processes involve electrons and ions, but electron–molecule collisions also give radicals in quite high concentrations. Polyatomic molecules yield a variety of radical fragments, but the method is more useful with diatomic molecules where the only radical species possible are the atoms. The main use of discharges in gas kinetics is the production of high atom concentrations by microwave or radiofrequency discharges, for example with hydrogen, oxygen or halogens.

2.3 Thermal Activation

This is perhaps the most familiar energy source—a thermostatted bath or furnace in which a reaction vessel is immersed. Molecules gain energy by collision with the vessel walls and this energy is distributed further by molecular collisions in the gas phase. The continuing process of energy transfer in molecular collisions maintains the equilibrium energy distributions for translational energy and internal energy, at the ambient temperature of the system. These distributions are described by equations due to Maxwell and Boltzmann. A molecule's translational energy ε is given by $\varepsilon = \frac{1}{2}mv^2$ where v is the molecular velocity*. The distribution of molecular velocities is:

$$\frac{\mathrm{d}N}{N} = 4\pi\left(\frac{m}{2\pi\bar{k}T}\right)^{3/2} \mathrm{e}^{-mv^2/2\bar{k}T}\, v^2\, \mathrm{d}v \tag{2.3.1}$$

where $\mathrm{d}N/N$ is the fraction of molecules with speed between v and $v + \mathrm{d}v$ regardless of direction. From this fundamental equation it is easy to derive the analogous distribution of translational energy and other useful functions such as the distribution of relative velocities for colliding molecules, the distribution of relative translational energies for colliding molecules, and average properties, e.g. average velocity and average relative velocity.

Typical velocity distributions are shown in figure 2.1 at two temperatures, illustrating the way equilibrium distribution is displaced to higher velocities as the temperature is increased. Translational energy distributions are similar in form to those shown in figure 2.1.

Internal energy distributions are conveniently represented by the ratio of number of molecules in two specific energy states, say N_2 in an excited state (ε_2) and N_1 in the lowest or ground state (ε_1). The Boltzmann equation gives:

$$(N_2/N_1) = \mathrm{e}^{-(\varepsilon_2-\varepsilon_1)/\bar{k}T} \tag{2.3.2}$$

* Throughout the book, ε will be used to denote molecular energy and E the energy per mole.

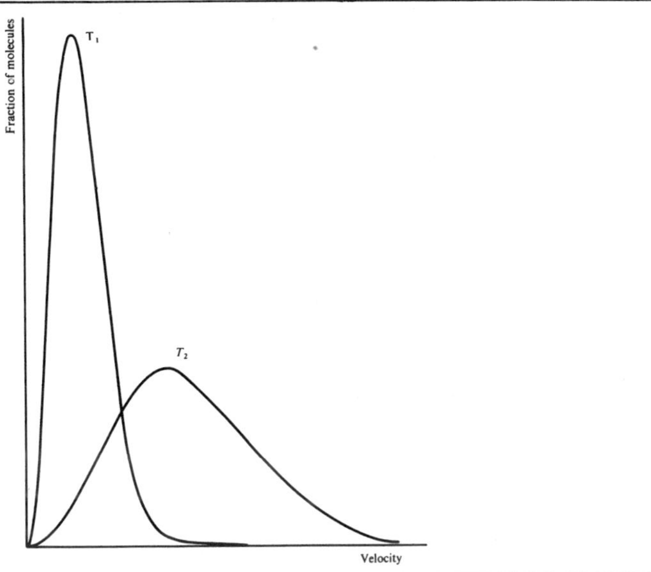

Figure 2.1
Maxwell–Boltzmann velocity distribution at two temperatures ($T_2 > T_1$).

The fraction of molecules possessing sufficient energy for a bond to break (e.g. as in reaction (2.3.I), below) must increase markedly with temperature:

$$C_2H_6 \rightleftharpoons 2CH_3 \tag{2.3.I}$$

As the equilibrium concentration of radicals increases, the overall rates of complex reactions in which they are involved increase. Many complex gas reactions with very slow rates at room temperature (~300 K) proceed at conveniently measurable rates in the temperature range 500–1 000 K which is easily accessible experimentally. A good example is the pyrolysis of ethane in which radicals are initially formed in reaction (2.3.I).

Conventional thermal experimental methods employ a reaction vessel immersed in a thermostatted furnace, with a suitable arrangement for measuring reactant and product concentrations at a series of times.

2.4 Photochemical Activation

Quanta of ultraviolet and visible radiation have energy in the range 150–600 kJ mol^{-1}, so when molecules absorb radiation from this region of the spectrum chemically important changes can occur. The molecules are excited to higher electronic states and eventually may dissociate.

In the primary photochemical act one quantum activates only one molecule (Einstein's *law of photochemical equivalence*). The rate at which quanta are absorbed is described by the empirical Beer–Lambert law. If the light passes through a vessel, length l, containing a substance at concentration c, and the incident light has intensity I_0 (quanta s^{-1}), then the intensity of light transmitted, I_t is given by:

$$\log_{10}\frac{I_0}{I_t} = \varepsilon(\lambda)cl \qquad (2.4.1)$$

where $\varepsilon(\lambda)$ is the extinction coefficient, or absorption coefficient. The value of $\varepsilon(\lambda)$ indicates how strongly the substance absorbs at wavelength λ. The intensity of light absorbed I_a is given by

$$I_a = I_0 - I_t$$

Absorbing a quantum of ultraviolet or visible radiation gives an electronically excited state of the molecule, and if the energy of the quantum is above the appropriate dissociation limit for the molecule dissociation follows rapidly. Some of the molecular processes leading to dissociation are discussed in more detail in sections 2.8–9 below, but for details of molecular spectra the reader should turn to a standard text on spectroscopy.

Photochemical dissociation may be represented:

$$M + h\nu \rightarrow A + B$$

and the rate of formation of radicals is given by

$$\mathrm{d}[A]/\mathrm{d}t = \mathrm{d}[B]/\mathrm{d}t = \text{const. } I_a$$

where the constant converts the rate to appropriate units.

2.5 Photosensitization

Not all molecules absorb radiation in an experimentally convenient region of the spectrum. In such cases it is sometimes possible to produce radicals by an indirect photochemical method using a *photosensitizer* which does absorb radiation at suitable wavelengths.

The photosensitizer is excited to an upper electronic state and this excitation energy can be transferred to the parent molecule on collision. An important example is mercury-sensitized decomposition of hydrogen.

$$\mathrm{Hg} + h\nu \rightarrow \mathrm{Hg}^*$$

$$\mathrm{Hg}^* + \mathrm{H}_2 \rightarrow \mathrm{Hg} + \mathrm{H}^{\cdot} + \mathrm{H}^{\cdot}$$

In another type of photosensitization, the sensitizer itself dissociates after absorbing radiation and the radicals so formed initiate further reactions in the main system. For example the reaction between hydrogen and oxygen can proceed at much lower temperatures when photosensitized by nitrogen dioxide, which absorbs light and dissociates:

$$NO_2 + h\nu \rightarrow NO + O^{\cdot\cdot}$$

The oxygen atom has two unpaired electrons and is written $O^{\cdot\cdot}$. The oxygen atoms then initiate the series of radical reactions in the hydrogen–oxygen system:

$$O^{\cdot\cdot} + H_2 \rightarrow OH^{\cdot} + H^{\cdot}$$

$$OH^{\cdot} + H_2 \rightarrow H_2O + H^{\cdot} \qquad \ldots \text{etc}$$

2.6 Usefulness of Photochemical Methods, Quantum Yield

Photochemical activation can only be employed where molecules have a sufficiently strong absorption spectrum in an experimentally accessible region, but it has the advantage that photochemical techniques are relatively simple. Reaction vessels are usually made of pyrex-type glass, which transmits wavelengths down to about 330 nm, or quartz which transmits down to around 180 nm when very pure. However, below about 200 nm oxygen in air absorbs radiation, so techniques of greater sophistication become necessary, and further complications ensue below 180 nm where the transparent materials used, such as calcium fluoride, are less easy to work with than quartz. Readily available light sources include gas discharge lamps emitting a continuous range of wavelengths in the ultraviolet and visible regions of the spectrum, and metal vapour arc lamps which emit intense lines characteristic of the metal. A comprehensive discussion of photochemical techniques may be found in the standard texts cited at the end of the chapter.

Despite these limitations, photochemical activation can, for a wide range of molecules, provide higher radical concentrations than found in thermal systems except at very high temperatures. Furthermore, the rate of radical formation depends on the radiation intensity and so can be controlled to some extent.

Combining various light sources with filters or a monochromator can limit incident radiation to a narrow band of frequencies or even a single spectral line. The use of approximately monochromatic radiation results in formation of approximately monoenergetic radicals, and this opens up new fields of kinetic studies where reactants do not have equilbrium energy distribution.

Measuring the amount of light absorbed by a photochemical system combined with chemical analysis of reactants and products can provide a valuable clue to the reaction mechanism. Actinometry, the measurement of light intensities, employs a variety of standard methods to give the total number of quanta absorbed during an experiment. The *quantum yield*, ϕ, is then defined:

$$\phi_{\text{product}} = \frac{\text{number of product molecules formed}}{\text{number of quanta absorbed}}$$

An alternative expression ϕ_{reactant} is defined by the number of reactant molecules removed. Since one quantum activates only one molecule in the primary photochemical process, ϕ gives the extent of reaction following each molecular activation—it establishes the number of secondary reactions that follow the primary process.

Various possibilities exist for primary photochemical processes depending on the types of molecular electronic states involved. The processes may lead to molecular excited states or dissociation to radicals, some of which may also be in excited energy states. A more detailed molecular picture of these possibilities is important to the kineticist even if it is displayed at a level more humble than necessary for spectroscopists or theoreticians. It is convenient to start with the simplest case—the diatomic molecule—and photochemical changes can be represented conveniently on potential energy diagrams for these molecules.

2.7 Potential Energy Diagram for a Diatomic Molecule

These diagrams show how a molecule's energy changes with the distance apart of the two atoms (the internuclear distance). To a good approximation the internal energy of a molecule is the sum of separate contributions for electronic, vibrational and rotational energy

$$E = E_{\text{el}} + E_{\text{vib}} + E_{\text{rot}}$$

Rotational quanta are much smaller than vibrational and electronic quanta, so rotational energy plays a comparatively small kinetic role and will be neglected in this discussion.

Consider first the potential energy curve for the ground electronic state of a molecule formed from atoms in their ground electronic states, as shown in figure 2.2. Species in higher electronic states would be represented by a similar diagram displaced to a higher energy overall.

High internuclear separation (high r) means the atoms are separate. As atoms come together, as r decreases, the attractive force between the atoms increases and energy falls. A repulsive interatomic force, which is negligible at large r, increases rapidly at small values of r and then the energy increases. Thus the energy of the system passes through a minimum at r_0 where the molecule would be in its most stable configuration: r_0 is the equilibrium *bond length.*

The effect of vibration can be expressed on this diagram of the ground electronic state. The lowest vibrational level ($v = 0$) does not coincide with the energy minimum but is raised above this by the *zero-point energy*. The $v = 0$ and several higher levels are shown in figure 2.2. This diagram also shows that as vibrational levels increase, the difference in energy between them decreases. This continues until the levels converge at the theoretical limit $v = \infty$ which is also indicated. On the classical picture based on Newtonian mechanics the diagram shows, for each vibrational level, the limiting values of interatomic distance between which the molecule vibrates, e.g. r' and r'' in the ground vibrational level. Quantum mechanics gives a quite different picture for lower vibrational levels. However, according to both views, as vibrational excitation increases the molecule vibrates over greater interatomic

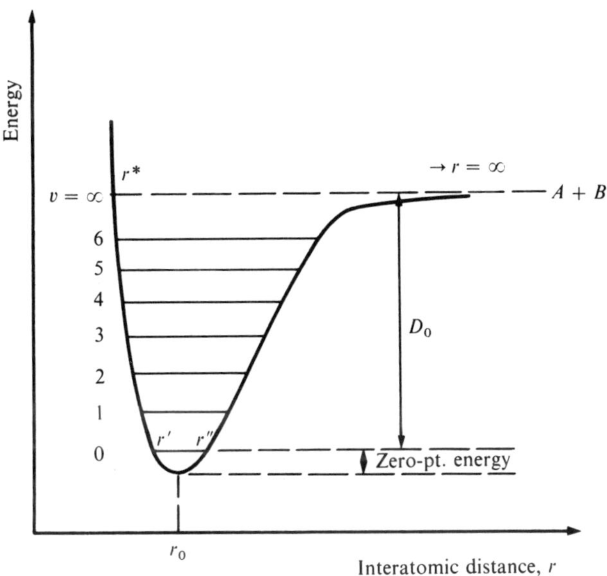

Figure 2.2
Potential energy diagram for the ground electronic state of a diatomic molecule.

distances—in simple terms the vibration becomes progressively more violent. In the $v = \infty$ level the limits to the vibration are r^* and infinity, so complete separation of atoms occurs within one vibration—in other words the molecule dissociates.

At normal temperatures and pressures virtually all molecules are in their ground electronic state and about 99·99% are in the ground vibrational state. The *bond dissociation energy* is the difference between the energy of the $v = 0$ level and that of separate atoms. This is shown as D_0 in figure 2.2.

2.8 Absorption of u.v. or Visible Radiation

Excited electronic states and fluorescence

Absorbing a quantum in this region of the spectrum excites a molecule to a higher electronic state in an *allowed transition*. 'Allowed' means obeying quantum mechanical rules such as those governing the relationship between the spins of the two electronic states. *Forbidden* transitions contravene these rules and occur with much lower probability. The ultimate fate of the absorbing molecule depends on the nature of the excited state and the energy of the absorbed quantum. First, consider the case shown in figure 2.3 which portrays the ground and first excited electronic state of a diatomic molecule with a few vibrational levels indicated. The energy of the absorbed quantum in the example is sufficient to excite the molecule to the $v' = 2$ vibrational level of the upper electronic state, and the transition is represented by the solid vertical line.

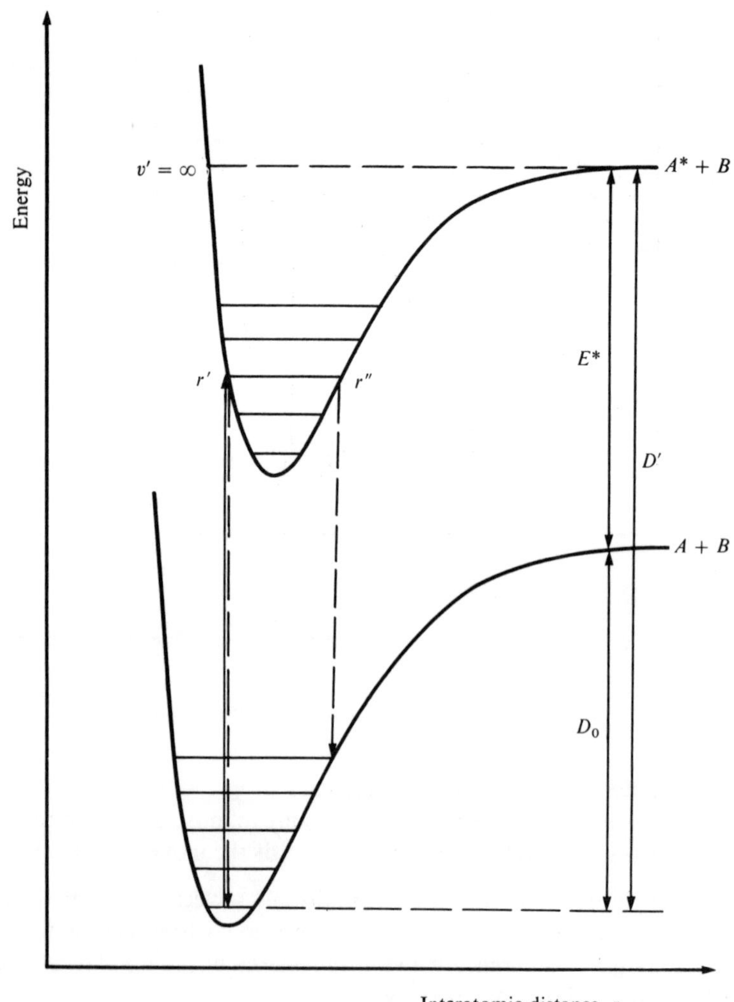

Figure 2.3
Ground and excited electronic states of a diatomic molecule showing absorption and emission (fluorescence) of radiation.

This transition occurs rapidly—in approximately 10^{-16} s. Since a typical vibrational period is longer than 10^{-14} s, the internuclear distance changes very little during the electronic transition. This justifies representing the transition, as in figure 2.3, by a change at constant r, i.e. by a vertical line. This construction is known as the *Franck–Condon* principle.

This primary process, electronic excitation, may be symbolized:

$$M + h\nu \rightarrow M^*$$

The most probable fate of the electronically excited molecule M^* is to return to the ground state in an allowed transition, the energy being emitted as a quantum of radiation. The lifetime of a typical electronically excited state is 10^{-8} s, and the transition may be to the ground vibrational level of the ground electronic state—the exact reverse of excitation:

$$M^* \rightarrow M + h\nu$$

This is not the only possibility for emission. According to quantum theory, in the lowest vibrational level the molecule spends most time around the mid-point of its vibration, but as vibrational quantum number increases the molecule spends more time at the extremes of the classical vibrational amplitude, e.g. r' and r'' for the level shown in figure 2.3. From higher vibrational levels the return to the ground state occurs most probably from either extreme, r' as indicated above, or r''. Both processes are shown by broken lines in figure 2.3. From r'' the molecule enters an excited vibrational level of the ground electronic state ($M^\dagger$) and a lower energy (longer wavelength) quantum is emitted:

$$M^* \rightarrow M^\dagger + h\nu'$$

A good example of this type of process is the formation of vibrationally excited nitric oxide. This is an example of the possibility, always present in photochemical studies, that excited energy states might be produced in higher concentrations than the equilibrium values.

The emission of radiation in allowed transitions is known as *fluorescence*.

Dissociation

If the molecule absorbs a quantum of high enough energy and enters the upper electronic state above the $v' = \infty$ level, then within one vibrational period the molecule will dissociate. This critical energy for the absorbed quantum is of magnitude D', as indicated on figure 2.3. If molecules absorb frequencies less than ν^* where $D' = h\nu^*$ the result is molecular excitation only, but frequencies greater than ν^* result in dissociation and these frequencies correspond to the continuous region in the molecule's absorption spectrum.

Dissociation from excited electronic states clearly requires more energy (D') than for dissociation from the ground state (D_0). The extra energy is shown as E^* on figure 2.3, and is due to electronic excitation of one of the atoms, A^* in the diagram. Thus the overall process can be symbolized:

$$M + h\nu^* \rightarrow A^* + B$$

A simplified version of this potential energy diagram is given in figure 2.4(a). Diatomic molecules may possess other types of excited electronic states and two other important examples are given in figures 2.4(b) and (c). In figure 2.4(b) the dissociation limit for the upper electronic state is the same as that of the ground state, and absorption of quanta with energy greater than D_0 leads to dissociation to ground state atoms:

$$M + h\nu \rightarrow A + B$$

The excited state shown in figure 2.4(c) is of the type called *repulsive*. Excitation to such states inevitably leads to dissociation, since the energy of the system decreases

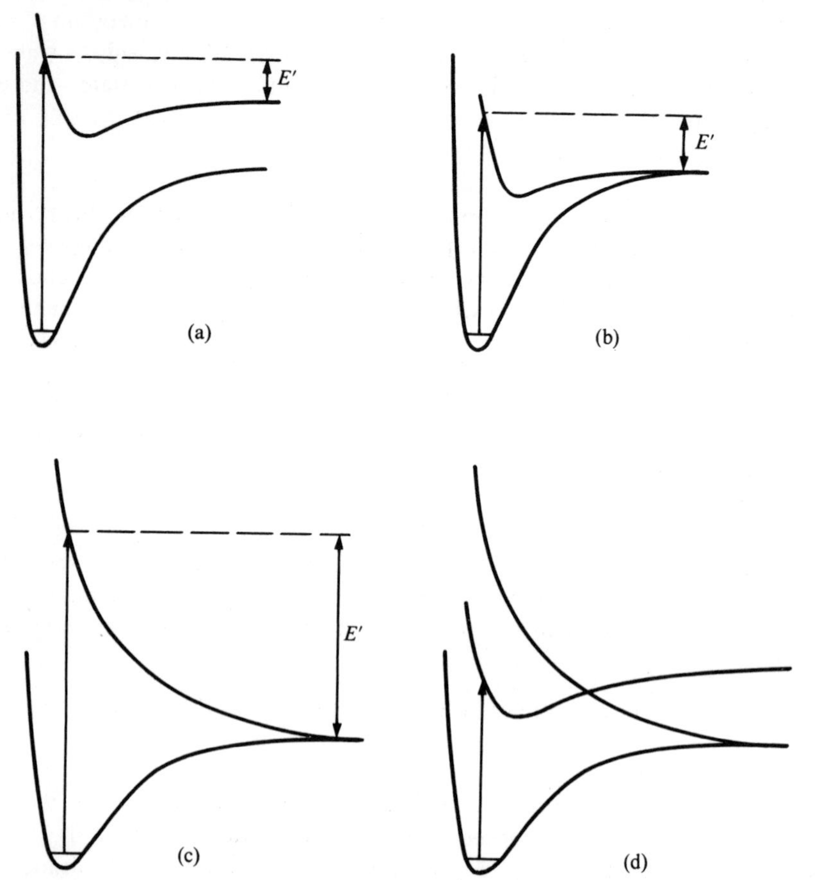

Figure 2.4
Possibilities for dissociation following absorption of radiation by diatomic molecules.

continually as r increases until the atoms are separate. In the example shown the dissociation limit is again the same as for the ground state, leading to ground state atoms.

Another feature of figures 2.4(a), (b) and (c) warrants comment. The size of the absorbed quantum, given by the vertical lines, is more than sufficient to cause dissociation in each case. There is excess energy marked E' still unaccounted for. Energy must be conserved when the diatomic molecule dissociates and this excess must appear as translational energy of the separating atoms. So, initially, photolysis may result in a non-equilibrium distribution of translational energy of the atoms.

Pre-dissociation: radiationless transitions

Molecules possess many excited electronic states and in certain cases the potential energy curves of excited states cross. However, theoretical rules 'forbid' such molecules from changing electronic states when the crossing point is reached during a vibration.

In practice the forbidden crossing does occur, though the probability is very low with diatomic molecules.

Such crossing can lead to dissociation even though the primary transition is to below the dissociation limit of the upper state. If the energy of the absorbed quantum is above that of the crossing point, as shown in figure 2.4(d), in each vibration there is a chance of transition to the repulsive state followed by inevitable dissociation. This is responsible for the phenomenon of *pre-dissociation* in molecular spectra. The crossing over from one excited state to another occurs at a single energy and is not accompanied by the emission or absorption of radiation. Such processes are referred to as *radiationless transitions* and are particularly important with certain polyatomic molecules (see section 2.9).

2.9 Polyatomic Molecules

In polyatomic molecules, absorbing a quantum in the u.v. or visible range results in excited molecular states or dissociation but the range of possibilities is greater. Dissociation will not, of course, give two atoms but a polyatomic radical and an atom, or two polyatomic radicals; e.g.:

$$NH_3 + h\nu \rightarrow NH_2^{\cdot} + H^{\cdot}$$

$$CH_3COCH_3 + h\nu \rightarrow CH_3CO^{\cdot} + CH_3^{\cdot}$$

Absorbed energy in excess of that required for dissociation will appear as internal and translational energy of the radicals.

Phosphorescence

Absorption of radiation by some molecules can give metastable electronically excited states, prominent examples being molecules with more than one aromatic ring, from naphthalene to biochemically interesting molecules.

In such cases the ground electronic state has all electrons paired in molecular orbitals to give a singlet state which may be symbolized S. Some excited states are also singlets (S*) while others have two unpaired electrons and are triplet states (T). A complete energy diagram for such molecules requires multidimensional space and a much simplified schematic diagram for a molecule with S* and T excited states is given in figure 2.5. Somc vibrational levels are portrayed in each state and high vibrational levels of the lower states are shown overlapping the low vibrational levels of other states.

When suitable quanta are absorbed the allowed transition:

$$S + h\nu \rightarrow S^*$$

takes place. The S* state can return to the ground state by more than one route:

(a) fluorescence, in the allowed transition:

$$S^* \rightarrow S + h\nu$$

(b) radiationless transition to a high vibrational level of the ground state

$$S^* \rightsquigarrow S$$

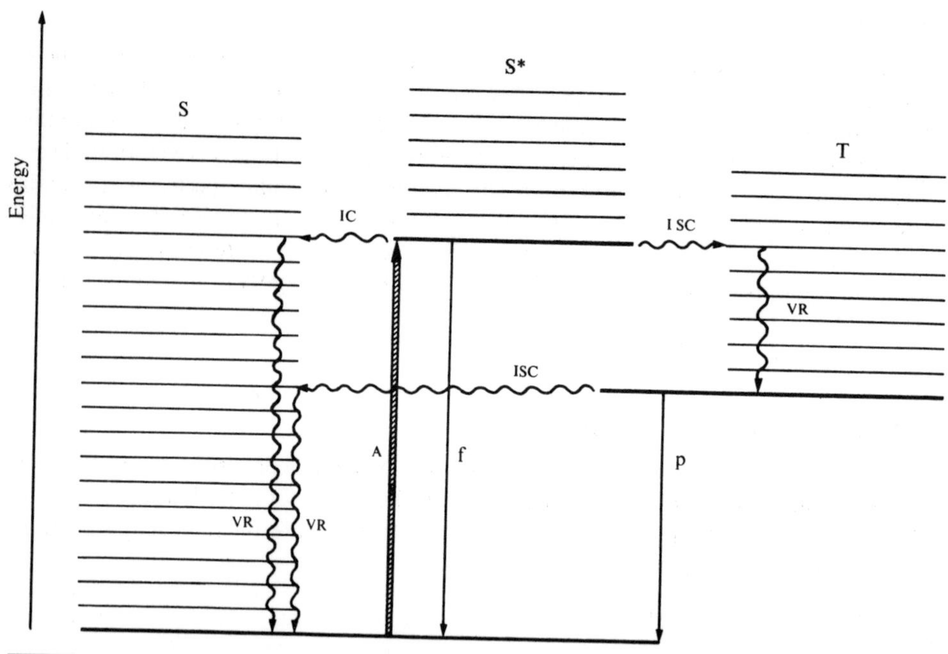

Figure 2.5
The population of triplet states by absorption of radiation (A) followed by intersystem crossing (ISC) and vibrational relaxation (VR). The excited singlet can also decay by fluorescence (f) or internal conversion (IC) and VR.

The triplet decays by phosphorescence (p) or reverse ISC and VR.

This is an example of the general process of radiationless transition to an electronic state of the same multiplicity, called *internal conversion* (IC), and it is illustrated in figure 2.5.

(c) via the triplet state.

The triplet state would be populated by a radiationless transition:

$$S^* \rightsquigarrow T$$

Such a transition to a state of different multiplicity is termed *intersystem crossing* (ISC) and is also illustrated in figure 2.5. Although this process is forbidden, selection rules apply to polyatomics much less stringently than to diatomics and a significant fraction of S* molecules follow this path.

In such systems, which are often studied in a condensed phase, molecules which enter new electronic states in high vibrational levels rapidly dissipate this excitation in collision with neighbouring molecules and so reach the lowest vibrational level of the state by vibrational relaxation (VR) as shown in figure 2.5.

There are two competing routes for the return of the triplet states to the ground state:

(i) The emission of radiation

$$T \rightarrow S + hv'$$

This forbidden process is relatively slow, typical lifetimes being in the range 10^{-3} s to several seconds for various molecules. This long-lived emission is thus clearly distinguished from the faster fluorescence and is called *phosphorescence**. The light is emitted at longer wavelengths than the fluorescence since phosphorescence occurs from the lowest vibrational level of the triplet.

(ii) Reverse intersystem crossing to a high vibrational level of the ground state

$$T \rightsquigarrow S$$

This is a slow process and is followed by rapid relaxation to low vibrational levels (VR).

The relative probability of triplet decay by phosphorescence or intersystem crossing varies with different molecules, but in all cases the triplet excited states are relatively long lived, *metastable*, and play important roles in the photochemistry of the molecules.

2.10 Excited States and Energy Transfer

Investigation of energy transfer between internal energy states of molecules and between internal and translational energy forms an important and rapidly expanding branch of gas kinetics. It embraces such techniques as fluorescence quenching, ultrasonic dispersion, molecular beams, laser-induced fluorescence spectroscopy and microwave–microwave double resonance. This field of study is not pursued at length in this text and the present section gives a brief summary of the salient features of energy transfer processes with some useful semi-quantitative comparisons[1].

Electronically, vibrationally and translationally excited species have appeared in several contexts in the preceding pages. Although rotational excitation has not been treated explicitly, it usually accompanies vibrational excitation. Such excited states can be produced by the absorption of radiation, by non-reactive molecular collision and as the products of chemical reaction. They may exist at concentrations higher than the equilibrium values for the ambient temperature of the system. The equilibrium distribution of translational energy was portrayed in figure 1.1, and table 2.1 gives the fractions of molecules in excited states for equilibrium at room temperature. Excitation may affect reactivity so it is important to consider the rate at which excited species decay, or relax, to their equilibrium concentration. There are two main energy transfer processes involved, the emission of radiation (fluorescence), and transfer in molecular collisions.

Competition between the two routes can be expressed:

$$A^* \xrightarrow{k_1} A + hv \qquad \text{fluorescence}$$

$$A^* + M \xrightarrow{k_2} A + M^* \qquad \text{collisional transfer}$$

* It is a historical oddity that the term derives from the emission observed when phosphorus is oxidized, but that process is an example of *chemiluminescence* (see below).

[1] A useful recent account of molecular energy transfer is given in R. D. Levine and R. B. Bernstein, *Molecular Reaction Dynamics*, Oxford University Press, 1974.

Table 2.1
Typical quanta and fractions of molecules in excited states for electronic, vibrational and rotational energy at S.T.P.

Internal energy	Typical quantum/ kJ mol^{-1}	Fraction in excited states at S.T.P.
electronic	300	$\sim 10^{-50}$
vibrational	20	$\sim 10^{-4}$
rotational	0·4	$\sim 0{\cdot}9$

Thus fluorescence is a first-order process with rate parameter k_1. It is also convenient to specify the relaxation time (equation (1.6.1), section 1.6) or fluorescent lifetime $\tau_1 = 1/k_1$. The more efficient the process, the larger k_1 and the smaller τ_1.

Collisional energy transfer may be regarded as a pseudo-first-order process. 'M' stands for any molecule with which the excited species can collide. It is thus a function of the bulk concentration $[M]$, proportional to total pressure. The rate of this step is given by $k_2[M][A^*]$ or $k_2'[A^*]$ for the pseudo-first-order process. The collisional lifetime is thus $T_2 = 1/k_2' = 1/k_2[M]$. For each type of energy the predominant process will be that with the shorter lifetime under the experimental conditions.

Electronically excited states

Theory shows that rate of fluorescence depends on ν^3 where ν is the frequency of the emitted quantum. For electronic quanta energy and frequency are high, so electronic fluorescence has a high rate and electronically excited states have very short lifetimes, about 10^{-8} s.

The frequency of molecular collisions in the gas phase can be calculated by the simple collision theory (section 5.1). At S.T.P. a single species undergoes a collision about every 10^{-10} s. However, the probability of electronic–translational energy transfer on collision, involving the sudden dissipation of such a large quantum, is very low. It occurs on average only once every 10^6 collisions, for example. Thus the average collisional lifetime is 10^{-4} s in this example, far longer than the fluorescent lifetime, and in general fluorescence is the dominant mode of decay for electronic excitation.

Collisional processes compete effectively with fluorescence only when the colliding molecule has internal energy states very similar in energy to the excited species. The energy level in question may be electronic or a high vibrational level of the ground state. In these cases *near-resonance transfer* takes place with much greater efficiency than for electronic–translational transfer. The probability of near-resonance transfer to a high vibrational level is greatest for polyatomic molecules. These have many vibrational modes, and there is a greater chance that a vibrational level almost exactly matches the electronic quantum.

When near-resonance transfer competes effectively with fluorescence it may be possible for an experimenter to increase total pressure until collisional transfer predominates. Fluorescence is thus suppressed, a phenomenon which is termed the *quenching of fluorescence*. If the fluorescent lifetime is known independently it is then possible to measure the rate parameters for collisional transfer.

Vibrationally excited states

Vibrational quanta are emitted in the infrared region of the spectrum, at much lower frequencies than electronic quanta, and typical fluorescent lifetimes for vibrational states are of the order of 10^{-3} s. For highly excited vibrational levels, efficient near-resonance collisional transfer may give rapid relaxation to the first excited level. For this level, transferring the comparatively small vibrational quantum to translational energy is much more efficient than for electronic quanta. Typically it may occur once in about 10^4 collisions, which, for 10^{10} collisions s^{-1} at S.T.P., means an average collisional lifetime of 10^{-6} s, much shorter than the fluorescent lifetime. Thus in general collisional energy transfer is the predominant mode of decay for vibrational excitation. Vibrational fluorescence can be observed by reducing the pressure (and so the relative probability of collision) to a very low value, usually less than 1 Torr, 133 N m^{-2}.

Rotational excitation

Rotational quanta are of very low energy and it is overwhelmingly probable that rotational excitation decays by collision. The rotational–translational transfer is very efficient and rotational energy is rapidly equilibrated.

Translational excitation

The exchange of translational energy on collision is the most efficient process considered here and translational excitation is rapidly equilibrated. Such translational–translational energy transfer is sometimes referred to as *elastic*, whereas translational–internal transfer is designated *inelastic*.

2.11 Excited Products of Chemical Reactions, Chemiluminescence

Electronic

Two examples of reaction products formed in metastable excited electronic states (where the transition to ground state is forbidden) are:

$$H^{\cdot} + NO \rightarrow HNO^*$$

$$CO + O^{\cdot\cdot} \rightarrow CO_2^*$$

Such species can return to the ground state by emission in the relatively long-lived forbidden process. Where the metastable excited state originates in a chemical reaction the emission is called *chemiluminescence*.

Vibrational

There are many reactions where a high fraction of energy released in the reaction (its exothermicity) appears as vibrational energy of products denoted by the superscript †, e.g.:

$$O^{\cdot\cdot} + NO_2 \rightarrow NO + O_2^{\dagger}$$

$$H^{\cdot} + Cl_2 \rightarrow Cl^{\cdot} + HCl^{\dagger}$$

As stated above the predominant mode of decay is collisional transfer. However, if such reactions are studied at very low pressure, relaxation may be mainly by emission (vibrational chemiluminescence). Chemiluminescence studies can provide detailed information about reactions on the individual molecular scale (see section 10.1).

2.12 Role of Excited States in 'Conventional' Kinetics

In recent years kinetic techniques have been developed for producing excited species in relatively high concentration and it is possible to observe their formation and decay directly. As stated previously, energy transfer has become a field of study in its own right. There is a similar growth in information on the effect of excitation on the reactivity of the species.

Several succeeding chapters are concerned with the immense subject of 'conventional' kinetic systems where complex reactions are investigated under conditions of approximate thermal equilibrium. In a high proportion of these studies the energy source is thermal. Large excess concentrations for excited states are rare and, except at very low pressures, excited states decay rapidly to their equilibrium concentrations. The assumption of thermal equilibrium is generally sound.

With photochemical energy sources great care must be exercised in interpreting data as rapid formation of excited energy states is more likely. Generally, in conventional photochemical systems, fluorescence and collisional decay occur significantly faster than the reactions of radical intermediates and thermal equilibrium is probably maintained approximately but such reactions must be interpreted with care, especially at low pressures. Special caution must be exercised for experiments at low pressure where collisional decay is relatively slow.

Further Reading

E. A. Moelwyn Hughes, *Physical Chemistry*, Pergamon, Oxford, 1957. (Kinetic Molecular Theory of Gases and Maxwell–Boltzmann Distribution).

E. A. Guggenheim, *Boltzmann Distribution Law*, Interscience, New York, 1963.

J. G. Calvert and J. N. Pitts, *Photochemistry*, Wiley, London, 1967.

R. P. Wayne, *Photochemistry*, Butterworths, London, 1970.

Comprehensive Chemical Kinetics, Vol. 3 (Eds C. H. Bamford and C. F. H. Tipper), Elsevier, 1969. (Excited States and Energy Transfer).

M.T.P. International Review of Science, Physical Chemistry Series One, Vol. 9 (Ed. J. C. Polanyi), Butterworths University Park Press, 1972. (Energy Transfer).

Exercises

2.1

The distribution of molecular velocities, $f(v)$, is given by equation (2.3.1), which may be written: $\mathrm{d}N/N = f(v)\,\mathrm{d}v$. The average molecular velocity is defined: $\bar{v} = \int_0^\infty f(v)v\,\mathrm{d}v$, and the average translational energy is $\bar{\varepsilon} = \int_0^\infty f(v)\tfrac{1}{2}mv^2\,\mathrm{d}v$.

Show that: $\bar{v} = (8\bar{k}T/\pi m)^{1/2}$ and $\bar{\varepsilon} = 3\bar{k}T/2$.

The following integrals are required:

$$\int_0^\infty \mathrm{e}^{-ax^2}x^3\,\mathrm{d}x = 1/2a^2 \qquad \int_0^\infty \mathrm{e}^{-ax^2}x^4\,\mathrm{d}x = 3\pi^{1/2}/8a^{5/2}.$$

Calculate the average velocities of helium and argon at 273 K and the average translational energy per mole at this temperature.

(Ans: $\bar{v}_{Ar} = 3{\cdot}8 \times 10^2$ m s^{-1}; $\bar{v}_{He} = 1{\cdot}2 \times 10^3$ m s^{-1}; $\bar{\varepsilon} = 3{\cdot}4$ kJ mol^{-1})

2.2

The first excited electronic state of a molecule is 300 kJ mol^{-1} above the ground state. In the ground electronic state the $v = 1$ and $v = 2$ vibrational levels are 22 and 42 kJ mol^{-1} above the $v = 0$ level. What fraction of molecules are in the excited electronic state and in the two vibrationally excited levels of the ground electronic state at 300 K and 1 000 K?

Ans:

	Upper electronic	$v = 2$	$v = 1$
at 300 K	$5{\cdot}8 \times 10^{-53}$	$4{\cdot}86 \times 10^{-8}$	$1{\cdot}46 \times 10^{-4}$
at 1 000 K	$2{\cdot}1 \times 10^{-16}$	$6{\cdot}35 \times 10^{-3}$	$7{\cdot}08 \times 10^{-2}$

2.3

The conversion, $A \rightarrow B$, was investigated photochemically in a reaction cell of 10 cm path length. A, at a pressure of 100 Torr, temperature 273 K, was irradiated for 10 s with monochromatic light, $\lambda = 254$ nm, of incident intensity 10^{15} quanta s^{-1}. Analysis established that the yield of B was $3{\cdot}8 \times 10^{-8}$ mol. The absorption coefficient for A at 254 nm was $0{\cdot}17$ mol^{-1} l cm^{-1} (the coefficient is defined in equation (2.4.1)– note the units employed).

What is the quantum yield for the reaction, and what does a value of this magnitude indicate?

(Ans: 100)

2.4

The bond dissociation energy of diatomic molecule AB is 235 kJ mol^{-1}. When AB absorbs 215 nm radiation it dissociates to atom A in its ground state and atom B in an excited electronic state with electronic energy, 61 kJ mol^{-1}.

If the ratio of atomic masses $B{:}A$ is $5{\cdot}52{:}1$ what is the translational energy of A after photolysis? Assume the translational energy of AB before photolysis is negligible and use the principle of conservation of momentum.

(Ans: 221 kJ mol^{-1})

2.5

A vessel, light path 10 cm, containing gas A at atmospheric pressure and 273 K was irradiated for 1 s with monochromatic radiation corresponding to the allowed transition from the ground electronic state to an excited state. The absorption coefficient was 5×10^3 mol^{-1} l cm^{-1}, and the intensity of incident light was 10^{14} quanta s^{-1}. $0{\cdot}01\,\%$ of the excited molecules underwent rapid radiationless transfer to a metastable excited state from which decay to the ground state occurred exclusively by the emission of radiation. The rate constant for this relaxation was $0{\cdot}1$ s^{-1}.

Calculate the approximate number of molecules in the metastable state 10 s after irradiation ceased.

(Ans: $3{\cdot}7 \times 10^9$)

2.6
A mixture of A, which absorbs incident radiation, with Q, which does not, is irradiated with monochromatic radiation which excites A to a higher electronic state A^*. A^* decays either by fluorescence or collisional de-activation with rate constants k_1 and k_2 respectively. The measured intensities of absorbed light and fluorescence are I_a and I_f respectively. Show that:

$$I_f/I_a = 1/\{1 + k_2(Q)/k_1\}.$$

At 273 K, in the absence of Q, it was found that the relaxation time for fluorescence was, $\tau_1 = 10^{-7}$ s. In the presence of 700 Torr of Q, $I_f = 0{\cdot}5\, I_a$. Calculate k_2.

(Ans: $2{\cdot}4 \times 10^{11}$ mol^{-1} cm^3 s^{-1})

Chapter 3: Experimental Methods: Slow Reactions

Since Bodenstein's pioneering work on hydrogen–halogen reactions in the early years of this century[1], overall rates have been measured for an enormous number of gas phase reactions. Increasingly sophisticated techniques have been applied to systems of academic or directly commercial interest. Although modern methods make it possible to focus on individual 'fast' elementary reactions, there is great scope for work on relatively 'slow' overall reactions and such research still proceeds apace. The subject of this chapter is investigating such slow complex reactions by 'conventional' kinetic methods.

The distinction between slow and fast in kinetics is not precisely defined. As a *rough* division, slow refers to reactions with half-lives of a few seconds and above—reactions which are suitable for study by conventional methods. Fast reactions have half-lives down to the picosecond range and require some advanced technique for direct investigation, but indirect deductions about their rates are often possible from results of conventional kinetic experiments.

Measuring overall rates leads to experimental rate equations, their reaction orders, and overall rate constants, and varying temperature gives overall experimental activation energies. Further vital mechanistic evidence is provided by detecting and identifying radicals in the system. From such data it is often possible to deduce a reasonable reaction mechanism and even to calculate rate constants and activation energies for some of the elementary radical reactions.

A suitable experimental system must provide an energy source for the reaction and an analytic method for measuring reactant and product concentrations at different

[1] For example, M. Bodenstein and S. C. Lind, *Z. Phys. Chem.*, 7, 168 (1907).

reaction times. Sources of energy for molecular activation were discussed in chapter 2, and thermal and photochemical activation are most commonly used. The more important analytic methods in current use will be described below followed by an outline of techniques for detecting radicals. In conclusion there will be a brief summary of the more important ways in which activation and analysis are combined for conventional kinetic experiments.

(A) Analytic Methods

3.1 Gas Liquid Chromatography (g.l.c.)

This technique has made a major impact on kinetics in the last twenty years. It has far more scope, flexibility and range of operation than previously available methods and it requires comparatively small samples for analysis.

The principle of the method is that the sample is injected into a flowing gas that sweeps it through a chromatographic column. The column is packed with an absorbent solid, or an inert solid coated with a liquid solvent. The various components in the sample differ in their solubility in the solvent, or strength of absorption on the solid, and so pass through the column at different rates. Hence the components emerge from the column and pass into a detector at different times. The detector records the emergence of each component and, when suitably calibrated, it measures their concentration. It is also possible to identify a component from the characteristic time at which it emerges from the column. Such is the variety of columns, column packings (absorbent solids and liquid solvents) and detectors that the method can cope with most reaction mixtures—even isomers can sometimes be separated and analysed. The sensitivity of g.l.c. is such that with a 1 cm^3 sample of gas at 1 Torr, 133 N m^{-2}, pressure, components present as only 1 % of the sample are readily detectable, corresponding to about 10^{14} molecules cm^{-3}. The method is suitable for stable molecules, but not for radicals.

A g.l. chromatograph is easily coupled to a reaction vessel by a gas sampling valve and the sample taken may be so small that concentrations in the reaction vessel are virtually undisturbed. Analysis times for each sample can be as small as a few seconds, so concentration measurements can be repeated at quite short intervals. An example of a field to which g.l.c. has made important contributions is the pyrolysis of hydrocarbons, discussed in section 6.7.

3.2 Mass Spectrometry

Mass spectrometers are extensively used to study reactions involving ions (see chapter 10), but in the present context only analysis of neutral molecules and radicals is considered.

The method is simple in principle. A sample is passed into the ion-source region of the instrument where sample molecules are ionized by bombardment with electrons of controlled energy. These ions, characteristic of their parent molecules, then enter the analysing region where ions with different charge/mass ratios are separated and detected. The ions are separated in two basic ways: in the first, ions are deflected in electrical and magnetic fields, the amount of deflection depending on charge/mass ratio; in a time-of-flight spectrometer, ions take different times to traverse a flight tube before reaching the detector. Relative concentrations at different charge/mass ratios are measured from the ion currents at the detector. The mass spectrum—the pattern of ions derived from a given parent molecule—is characteristic and it is often possible to identify components in a mixture.

Very small samples are required, and they may be leaked directly from a reaction vessel, but for straightforward chemical analysis of complex mixtures there is a disadvantage. The mass spectrum of a compound may be complicated, with many different fragment ions resulting in the pattern referred to above. Where several compounds are present, parts of their patterns may coincide making results difficult to interpret. If this complication is absent, the method is sufficiently sensitive to detect approximately 10^{11} molecules cm^{-3}.

An area in which mass spectrometry is more successful than g.l.c. is isotopic analysis. A very useful trick for elucidating reaction mechanisms consists of isotopically labelling reactants and subsequently identifying isotopically mixed products. For this mass spectrometry is extremely convenient, due to the excellent resolution of small mass differences. The spectrometer may also be used simply as a detector in conjunction with g.l.c. separation.

Mass spectrometry is capable of identifying radicals in reaction mixtures. A sample leaked directly from the reaction vessel into the ion-source region suddenly enters a system at a pressure of around 10^{-4} Torr, $1{\cdot}3 \times 10^{-2}$ N m^{-2}, dropping further to about 10^{-8} Torr, $1{\cdot}3 \times 10^{-6}$ N m^{-2}, in the analysis region. At such pressures radicals undergo very few collisions with reactant molecules and they may have lifetimes long enough for analysis. The radicals often give the same fragment ions as a related molecule, but the ions from radicals appear at lower energies of bombarding electrons. This makes it possible to distinguish the radical mass spectrum from that of stable molecules. An important contribution to combustion kinetics was proving that the HO_2 radical did exist[2]. Its reactions are referred to frequently in chapter 7.

3.3 Absorption Spectroscopy

Ultraviolet, visible or even infrared spectra can be employed for following changes in concentration of reactants and products if the absorption is sufficiently strong and does not overlap the spectrum of another molecule present. Background radiation from a suitable light source is passed through the reaction vessel into a monochromator, which isolates the spectral region where the molecule absorbs. The intensity of transmitted light is measured photoelectrically and the photoelectric current can be recorded continuously. Thus changes in absorption in the selected wavelength range, and so changes in concentration can be recorded. Typical examples of molecules whose reactions may be followed this way are halogens and nitrogen dioxide.

In conventional kinetic systems the radical concentrations are generally below the limits for spectroscopic detectability. Spectroscopic analysis is widely used, however,

[2] A. J. B. Robertson, *Trans. Faraday Soc.*, **48**, 228 (1952); *Mass Spectrometry*, Methuen, London, 1954; S. N. Foner and R. L. Hudson, *J. Chem. Phys.*, **21**, 1374 (1953).

in conjunction with some modern techniques for producing high concentrations of radicals and investigating their reactions (see chapter 4).

(B) Detection of Radicals

Methods described in this section are of diagnostic importance in conventional systems, where just detecting and identifying a radical gives a valuable clue to the reaction mechanism. Several of the methods have been extensively employed for following radical concentrations in modern techniques where much higher concentrations are produced. These are discussed in chapter 4.

3.4 Metallic Mirrors

This historic method was first used by Paneth[3] nearly fifty years ago to demonstrate the existence of alkyl radicals in hydrocarbon reactions. A hydrocarbon was pyrolysed by flowing through a furnace and the resulting mixture passed over a thin layer, a mirror, of metal such as lead which had been deposited prior to the experiment. The mirror disappeared and on trapping the products they were found to contain metal alkyls, PbR_4. Thus the hydrocarbon reaction mixture must have contained alkyl radicals which removed the mirror:

$$Pb + 4R^{\cdot} \rightarrow PbR_4$$

3.5 Optical Spectroscopy

Emission spectra of radicals are readily observed in reactions carried out at very high temperature[4], where radicals are present in excited electronic states—in explosions or flames for example. It is difficult to extract more than diagnostic information but the list of species so detected is very long, some important examples being OH, CH, CN.

As stated above, use of absorption spectra in conventional systems is limited by low radical concentrations.

3.6 Atomic Resonance Radiation

This is a very recent development that is proving extremely useful in gas kinetics[5]. Intense emission of atomic resonance lines from excited atoms is obtained from microwave electrical discharges through gases and vapours. Typical emission sources include H, O, S, Pb and halogen atoms. This emission can be used in two basic ways to measure very low concentrations of the corresponding atoms in gas systems.

[3] E.g., F. Paneth and W. Hofeditz, *Ber. Dt. Chem. Gess.* **B62**, 1335 (1929).
[4] A. G. Gaydon, *Spectroscopy and Combustion Theory*, Chapman and Hall, London, 1948.
[5] A. A. Westenberg, *Ann. Rev. Phys. Chem.*, **24**, 78 (1973).

Attenuation of resonance radiation

The atomic radiation is passed through the reaction vessel to a photoelectric detector and when the vessel contains the same atoms as the emission source the resonance radiation is absorbed and the fall in intensity at the detector is measured. From this attenuation of the incident radiation the concentration of atoms in the vessel is easily calculated.

Atomic resonance fluorescence

Here ultimate detection limits of 10^{10} atoms cm^{-3} are reached, beyond the sensitivity of normal mass spectrometry or electron spin resonance. The reaction vessel is irradiated by the atomic resonance source, and atoms in the vessel absorb the resonance radiation as described above. The atoms which absorb the resonance radiation are, of course, excited to a higher electronic state, and they return to the ground state by fluorescence, emitting the same resonance radiation. This fluorescence can be observed at right angles to the direction of incident resonance radiation, and very low intensities can now be recorded photoelectrically.

3.7 Gas Titration

This is an analogue of the familiar titration in solution. The 'titrant' gas is added to the reaction mixture, and a critical change in the emission spectrum from the system provides the 'indicator'[6]. For example, the concentration of oxygen atoms can be measured by adding nitrogen dioxide as titrant. The titration reaction is:

$$O^{\cdot\cdot} + NO_2 \rightarrow NO + O_2$$

and the indicator is emission by chemiluminescence from excited nitrogen dioxide formed in the process:

$$O^{\cdot\cdot} + NO \rightarrow NO_2{}^*$$

Chemiluminescence occurs as long as oxygen atoms are present. Once the 'equivalent' amount of nitrogen dioxide has been added, i.e. once the end-point is reached, all oxygen atoms are removed. Hence the reactions producing $NO_2{}^*$ cease and the yellow–green chemiluminescence is extinguished. The method has been developed to sensitivities of about 10^{12} atoms cm^{-3}.

3.8 Electron Spin Resonance (e.s.r.)

Due to their unpaired electrons radicals are paramagnetic and the interaction of the electron spin with an applied magnetic field gives splitting of energy levels. This can be detected by resonance absorption of radiation in the microwave region, and atoms and radicals have characteristic electron spin resonance spectra[7] (sometimes called electron paramagnetic resonance spectra). Measured absorption intensities give the relative atom and radical concentrations in the system, but by direct calibration, using

[6] B. A. Thrush, *Science*, **470**, 156 (1967).
[7] A. Carrington, *Chemistry in Britain*, **4**, 301 (1968).

known pressures of oxygen or nitric oxide, these can be converted to absolute concentrations.

E.s.r. has been used for only ten years in direct studies of elementary radical reactions, but sensitivities of around 10^{11} species cm^{-3} have been attained and this, combined with reliability, has resulted in widespread use. Structural investigations of radicals have also been carried out by trapping them in a rapidly cooled solvent and studying detailed e.s.r. spectra of the radicals in the inert matrix of the solid solvent.

3.9 Isotopic Mixing and Mass Spectrometry

The detection of radicals by mass spectrometry was discussed in section 3.2, and its use to study isotopic mixing was also mentioned there. This method for demonstrating the participation of specific radicals in a reaction can be illustrated by the decomposition of dimethyl ether where an important product is methane, with the methyl radical as its precursor: $CH_3 + CH_3OCH_3 \rightarrow CH_4 + CH_2OCH_3$.

When nitric oxide is added as inhibitor, the residual rate was considered by many workers to represent a molecular rather than a radical reaction. However, pyrolysing a mixture of CH_3OCH_3 and CD_3OCD_3 gave the same relative yields of isotopically mixed methanes (CH_4, CH_3D, CD_3H, CD_4) with or without nitric oxide, confirming a radical mechanism in each case[8].

3.10 Calorimetry

If probes consisting of coated resistance wires are inserted into a reaction vessel, radicals re-combine on the probe surface and the reaction exothermicity results in a rise in temperature for the probe. This rise in temperature, appropriately calibrated, measures radical concentrations in the gas phase. The method has been used, for example, with halogen atoms, but it is non-specific and interferes with the concentrations of radicals measured.

3.11 Radical Scavenging

This is another method that was more popular in the past. If detection of radicals by the methods listed above is difficult in a given system, it may be achieved by 'trapping' a radical in a chemical reaction and identifying it from the resulting stable product. Such additives, for deliberately removing radicals by reaction, are called *scavengers*. For example, if hydrocarbon radicals (R) participate in a reaction and a halogen (X_2) is added, the radicals are scavenged in the very fast reaction:

$$R^{\cdot} + X_2 \rightarrow RX + X^{\cdot}$$

The stable halide can be analysed by g.l.c. and R identified. The method, of course, drastically interferes with the reaction and care must be exercised in mechanistic deductions.

[8] D. J. McKenney, B. W. Wojciechowski and K. J. Laidler, *Can. J. Chem.*, **41**, 1993 (1963).

(C) Conventional Kinetic Systems

Over the last seventy years a major body of kinetic data has been accumulated by methods which essentially consist of: (a) using high-vacuum techniques to prepare reactant mixtures at known concentrations; (b) heating or photolysing the mixture; (c) following the variation in reactant and product concentrations with time by appropriate analytic methods. Reaction may take place in a closed reaction vessel giving, in this sense, a *static system*, or in a *flow system* where analysis is carried out at the end of a flow tube.

3.12 Static Systems

Vacuum apparatus and pressure measurement

Typical reactant pressures are in the range 100–500 Torr, $1{\cdot}3 \times 10^4$–$6{\cdot}7 \times 10^4$ N m^{-2}. To eliminate impurities during preparation and mixing of reactants, operations are carried out on a high-vacuum line to which all components of the apparatus are attached. The line is evacuated by a combination of oil rotary pump and mercury or oil diffusion pump with liquid nitrogen traps. Background pressures of 10^{-5} Torr,

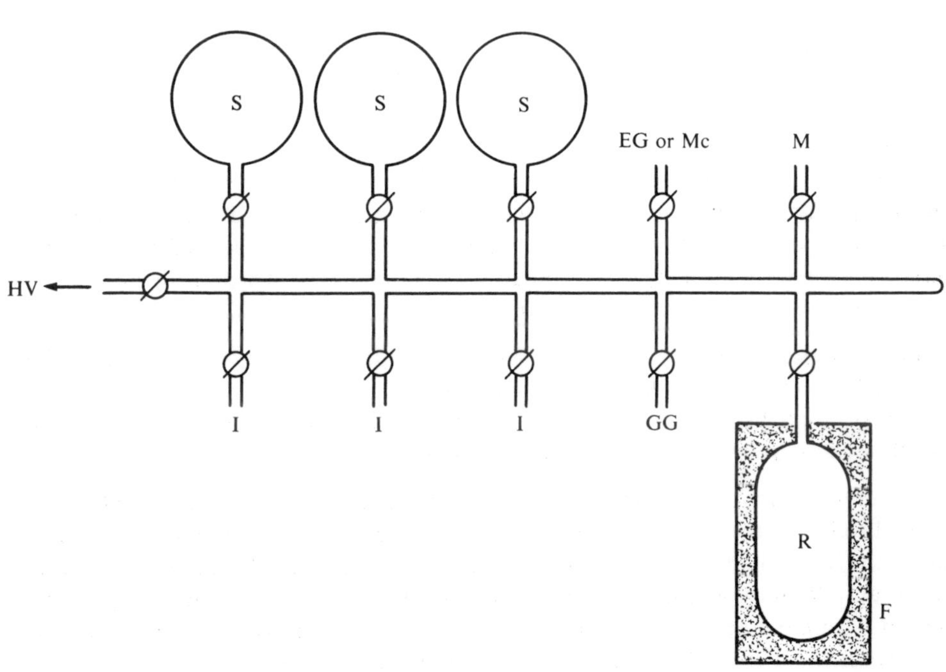

Figure 3.1
Diagrammatic representation of conventional static system: HV, to liquid N_2 traps, diffusion pump and rotary oil pump; S, gas storage vessels; I, inlets for gases; EG, electrical gauge; Mc, Mcleod gauge; GG, glass gauge; M, manometer; F, thermostatted furnace; R, reaction vessel. R may also have another arm for direct sampling to g.l.c., or F and R may have suitable windows for spectroscopic measurement.

$1{\cdot}3 \times 10^{-3}$ N m^{-2}, are reached easily and are measured with electrical gauges or McLeod mercury gauges. Reactant pressures are best measured with sensitive all glass gauges such as the spoon and spiral type, although if no corrosive gases are present mercury manometers are convenient for higher pressures, and oil manometers may be used for lower pressures.

Reaction vessel

The cylindrical or spherical vessel, probably a few hundred cubic centimetres in volume, is usually made of pyrex glass or quartz. Glass withstands temperatures up to about 800 K, and for photochemical studies is suitable for wavelengths down to around 350 nm where it starts to absorb strongly. Quartz may be used up to about 1300 K, and it transmits radiation at wavelengths down to around 200 nm, or about 180 nm if the quartz is very pure.

Steady high temperatures are readily provided by thermostatted furnaces in which the vessel is placed. An associated experimental problem is the 'dead space', some volume containing reactants but not exposed to the full reaction temperature, e.g. the stopcock and narrow tube leading to the vessel.

Reactants may be mixed in the vessel, or pre-mixed. The mixing time and the period required to reach the desired temperature are major factors which limit the reaction times that can be followed. At best, the time for mixing and temperature equilibration may be as low as one second, suitable for reactions with half-lives greater than ten seconds.

Analysis

In a kinetic experiment there may be continuous or intermittent analysis or even just one analysis at a chosen time. The single determination is least desirable. It usually means stopping the reaction at a certain time, for example by freezing the reaction vessel in liquid nitrogen, and then analysing the contents. Then there must be a separate experiment for each reaction time.

With intermittent measurement, samples are withdrawn at a series of times, the interval between samples allowing sufficient time for the analysis of each. G.l.c. is particularly suited to intermittent work since it can give accurate and almost complete sample analysis at intervals as low as a few seconds.

For reactions in which the number of moles changes, an early and popular continuous method followed the change in total pressure. However, measuring only one overall parameter for a complex reaction gives very limited information. Continuous monitoring of all relevant concentrations is rarely possible but some reactants or products can often be monitored by their absorption spectra, or by mass spectrometry with a continuous small 'leak' from the reaction vessel to the spectrometer.

The conventional static system briefly outlined above is represented schematically in figure 3.1. It has been used to measure reaction rates and so deduce mechanisms and kinetic parameters for the majority of important complex gas reactions. The results of such kinetic experiments are discussed in chapters 6 and 7 where examples include hydrogen–halogen reactions and the pyrolysis and combustion of hydrocarbons.

3.13 Flow Systems

Two main types can be distinguished, linear flow and stirred flow. Linear flow methods can cope with slightly faster reactions and the measurement of smaller concentrations than static methods. Often, for complex reactions, the results are not expressed readily as rate equations of the familiar type which apply to the constant-volume conditions found in static systems. This difficulty can be overcome by stirred flow techniques where rapid stirring provides a steady state in the mixture that leaves the vessel. A suitable ratio of stirring rate to reaction rate must be maintained and this limits the speed of reactions that can be studied.

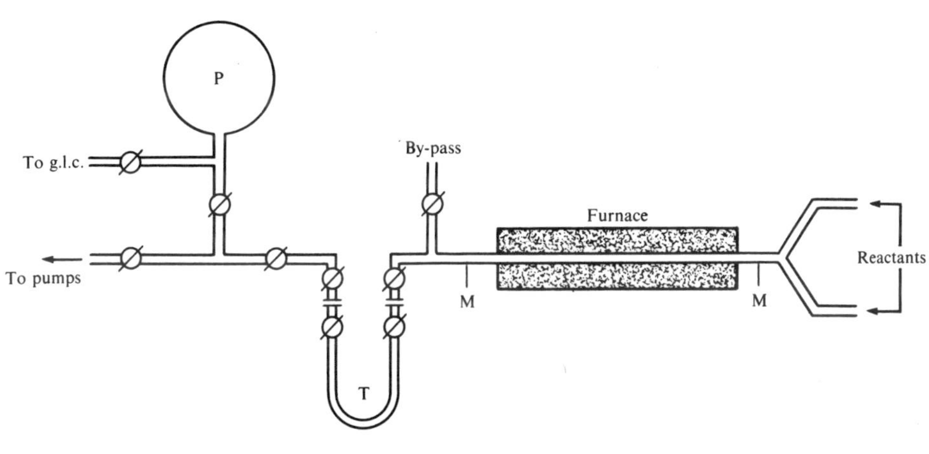

Figure 3.2
Diagrammatic representation of linear flow system: M, manometers; T, trap for liquid products (to be immersed in coolant); P, bulb for permanent gases.

Linear flow

Reactants are mixed and then flow at a constant, measured rate through a tube after which the mixture is analysed. The reaction time, t, is the time between mixing and leaving the tube, and this can be changed by varying the flow rate. If the length of the tube is l and the linear flow rate is u, $t = l/u$. The flow of gases is itself a complicated phenomenon but the desired viscous flow can be maintained over a wide range of experimental conditions.

Typical flow tubes are 2 cm in diameter and 30 cm long, and with gas pressures of 5–50 Torr, $6{\cdot}7 \times 10^2$–$6{\cdot}7 \times 10^3$ N m^{-2}, reaction times down to about 1 s can be attained. Once steady conditions are set up the emerging mixture is collected for analysis almost at leisure. With smaller concentrations, longer periods for collection are required and very small degrees of reaction can be studied. A schematic representation of a flow line is given in figure 3.2.

The linear flow method is most suitable for reactions with simple kinetics. The situation is further obscured if the reaction brings about a change in volume which is appreciable relative to the total volume of reactants. The linear flow method has been most popular for first-order or pseudo-first-order reactions, for example, the decomposition of toluene[9] to give hydrogen, methane, and benzene where the initial reaction is probably:

$$C_6H_5CH_3 \rightarrow C_6H_5CH_2^{\cdot} + H^{\cdot}$$

Stirred flow

Here the reactants flow into a vessel designed to achieve rapid uniform mixing by ensuring vigorous convection and diffusion of the reactants. When a steady state is reached the mixture flowing from the vessel is collected for analysis. Species in the vessel are at stationary concentration, and the rate at which each product emerges is equal to the rate at which it is formed. The assumption of complete mixing is substantiated better than for linear flow and kinetic parameters are derived more simply and reliably. A typical example of a reaction studied by this method is the decomposition of di-tertiary-butylperoxide[10], a first-order radical reaction.

Further Reading

H. W. Melville and B. G. Gowenlock, *Experimental Methods in Gas Reactions*, Macmillan, London, 1964.

Comprehensive Chemical Kinetics, Vol. 1, (Eds C. H. Bamford and C. F. H. Tipper), Elsevier, 1969.

Techniques of Chemistry, Vol. 6, Part 1, 3rd ed. (Ed. E. S. Lewis), Wiley Interscience, 1974.

Exercises

3.1

X atoms can be generated from X_2 by heat or photolysis. The rate constants k_1 and k_2

$$X_2 + X_2 \underset{k_2}{\overset{k_1}{\rightleftharpoons}} X + X + X_2$$

have the values, $k_1 = 10^{15}\, e^{-E_1/RT}$ mol^{-1} cm^3 s^{-1}, where $E_1 = 200$ kJ mol^{-1}, and $k_2 = 10^{15}$ mol^{-2} cm^6 s^{-1}. Calculate the equilibrium concentration of X when $[X_2] = 10^{-5}$ mol cm^{-3} at: (a) 300 K; and (b) 1 000 K.

X_2 is contained in a vessel 500 cm^3 in volume with a 30 cm light path at 300 K. It is irradiated with light of incident intensity 10^{15} quanta s^{-1} at a wavelength where the absorption coefficient (ε) of X_2 is 10^2 mol^{-1} l cm^{-1} and absorption leads to dissociation. Calculate the equilibrium concentration of X (c).

What is the significance of these results for kinetic investigations of reactions involving X_2?

(Ans: (a) $1{\cdot}2 \times 10^{-20}$; (b) $1{\cdot}9 \times 10^{-8}$; (c) $2{\cdot}6 \times 10^{-11}$ mol cm^{-3})

[9] S. J. Price, *Can. J. Chem.*, **40**, 1310 (1962).
[10] M. F. R. Mulcahy and D. J. Williams, *Aust. J. Chem.*, **14**, 534 (1961).

3.2
A deuterium halide, DX, on photolysis with suitable monochromatic radiation dissociates to give deuterium atoms of high translational energy. In DX, H_2 mixtures the excited atom D* may react as in (a) or (b) below

(a) $D^* + DX \rightarrow D_2 + X$

(b) $D^* + H_2 \rightarrow HD + H$

or it may lose energy in non-reactive collisions with either DX or H_2. Such thermalized D atoms are *all* scavenged by DX to give D_2. The D_2 and HD concentrations are easily measured by mass spectrometry. Show that the relative probability of processes (a) and (b) can be measured from a plot of product ratio versus reactant ratio, i.e. $[D_2]/[HD]$ versus $[DX]/[H_2]$.

(The answer is given in section 10.7)

3.3
The decomposition of ethane yields mainly ethylene and hydrogen and it is inhibited by the addition of nitric oxide. It has been suggested that the residual reaction rate in the presence of nitric oxide is due to a molecular reaction rather than radical reactions. How would you investigate this hypothesis if the various experimental techniques described in this chapter were at your command?

3.4
The reaction $A \rightarrow B$ is kinetically first order. When studied in a flow tube, length 30 cm, at linear flow rate 10 cm s^{-1} the product yield shows that 1% decomposition occurs. Calculate the first-order rate constant.

(Ans: $3{\cdot}3 \times 10^{-3}$ s^{-1})

3.5
The gas reaction of stoichiometry: $A + B \rightarrow C + D$, shows the following variation of reaction rate with partial pressures of the species:

Pressure/Torr				Rate, $-d[B]/dt/10^{-2}$ Torr s^{-1}
A	B	C	D	
100	100	0	0	10
200	100	0	0	20
100	200	0	0	10
100	100	0	100	10
100	100	100	0	5
100	100	400	0	2
100	100	100	100	5
400	100	100	0	20
100	400	100	0	8

By careful inspection of the data, empirically deduce the experimental rate equation.

Show that the mechanism given in exercise 1.9 corresponds to this rate equation where $k_2 = k_3$ and $k_1 = 10^{-3}$ s^{-1}.

3.6

When the thermal decomposition of ethane is studied in a static system the reaction stoichiometry in its early stages is well represented by: $C_2H_6 \rightarrow C_2H_4 + H_2$ with minor yields of CH_4 and C_4H_{10} also present. Early kineticists followed the change in total pressure and established first-order kinetics for the removal of ethane. Today the change in concentration of C_2H_4, CH_4, C_4H_{10} can be studied by g.l.c. Look at the mechanism given in section 6.7 and with the benefit of this hindsight compare the kinetic information that early and modern methods might yield, noting the pitfalls facing the interpretation of the early results.

Chapter 4: Experimental Methods: Fast Reactions

There is no precise definition of a fast reaction. Referring to half-lives is not satisfactory since, except for first-order reactions, the half-life varies with initial concentration. A convenient rule of thumb is to consider as fast those reactions for which the experimental methods described in chapter 3 are inadequate—broadly speaking where reaction times are less than a few seconds. Two general types of experimental difficulty arise: defining initial times; and achieving suitable time resolution when measuring concentration. For kinetic experiments, reactants are mixed together to start the reaction, or a reactive species is suddenly formed in a system previously at equilibrium. In both cases the starting processes take a finite time, and for a satisfactory kinetic investigation the mixing time or initiation time must be much shorter than the half-life of the reaction. After the initial time, the reaction is followed by measuring concentrations at a series of times. When these times are fractions of a second, kinetics enters a new technological world, one in which time resolution has gradually been refined from seconds to milliseconds, to microseconds, to nanoseconds and recently to picoseconds.

The last twenty-five years has seen the development of a range of techniques which make it possible to investigate directly virtually every type of elementary radical reaction and molecular energy transfer process. The impetus to these advances was provided by the development of electronic measuring devices in World War II, and experimental sophistication continues to grow. The kineticist, like other scientists, is now in an Electronic Age. Each kinetic technique must combine forming radicals or excited states, a subject discussed in chapter 2, and following their concentrations through their brief lifetimes by some variant of the detection methods discussed in sections 3.5–10.

In a text of this type and length it is not possible to provide an exhaustive survey of this field but the majority of important relevant techniques are discussed. Schemes for classifying these techniques have been devised, though demarcation is not meant to be rigid. One such scheme is to distinguish Perturbation, Competition, and Flow methods. In the first, a system originally at equilibrium is subjected to a sudden perturbation whereupon its rapid re-adjustment to a new equilibrium is followed. *Large perturbations* are provided by: intense pulses of light in *flash photolysis* and *pulsed laser* methods; by a pulse of high energy electrons in *pulse radiolysis*; and by thermal heating with a supersonic shock wave in *shock tubes*. The large perturbations produce very high concentrations of short-lived radicals or excited molecules which are usually observed by spectroscopic techniques. *Small perturbations* are employed in the wide range of *chemical relaxation methods*, where a small displacement is produced by temperature jump, pressure jump, electric field jump or ultrasonics, and then relaxation to the new equilibrium position is traced. All the large perturbation methods are used in gas phase studies but small perturbation methods are more appropriate for solution kinetics, with the exception of ultrasonic dispersion which is a major source of data on vibrational–translational energy transfer in gases.

Competition methods include observing line-broadening in spectra (especially n.m.r.) and quenching of fluorescence. The latter refers to competition between the decay of molecular excited states by fluorescence, and decay by de-activation in molecular collisions. *Flow methods* represent an extension for methods in which reactants are mixed, since in flow systems mixing times can be greatly reduced. High radical concentrations may be generated by electrical discharge through a gas which is then mixed with other reactants in a fast-flow system. For solution kinetics, further refinements such as accelerated or stopped flow have been developed.

This chapter will concentrate on techniques which have been fruitful in gas kinetics, or which seem to offer great future potential. Flash photolysis, pulsed lasers, shock tubes and fast flow are discussed at some length, pulse radiolysis only briefly. The exciting new field of *molecular beam* kinetics also deals with fast reactions, but this approach is so fundamentally different that it is described separately in chapter 9, and some applications of chemical lasers are outlined in chapter 10.

4.1 Flash Photolysis

The technique, developed by Norrish and Porter around 1950[1], employs an intense flash of light to produce transient species, radicals or excited molecules, in concentrations that may be four or five orders of magnitude greater than found in conventional systems. A flash of a few microseconds duration allows kinetic studies in the microsecond–millisecond region which embraces a wide range of radical reactions and energy transfer processes.

A typical apparatus is represented schematically in figure 4.1. The quartz reaction vessel is placed parallel to the initiating flash or 'photoflash', a quartz discharge tube containing an inert gas at low pressure. The condenser, C_1, is charged to a high

[1] R. G. W. Norrish and G. Porter, *Nature*, **164**, 658 (1949); G. Porter, *Proc. R. Soc.* (*London*), **A200**, 284 (1950).

voltage and when a trigger pulse (not shown) is applied the energy, hundreds or thousands of joules, discharges in a few microseconds producing an intense light flash which has the same duration. This radiation is emitted in a continuous range of wavelengths from the lowest transmitted by quartz (~200 nm) throughout the ultraviolet and visible, thus covering much of the photochemically useful range. To increase the amount of radiation absorbed, photoflash and reaction vessel are enclosed in a cylindrical reflector (not shown on the diagram). Radical concentrations of 10^{-8}–10^{-7} mol cm^{-3}, corresponding to pressures of about 1 Torr, 133 N m^{-2}, can be attained.

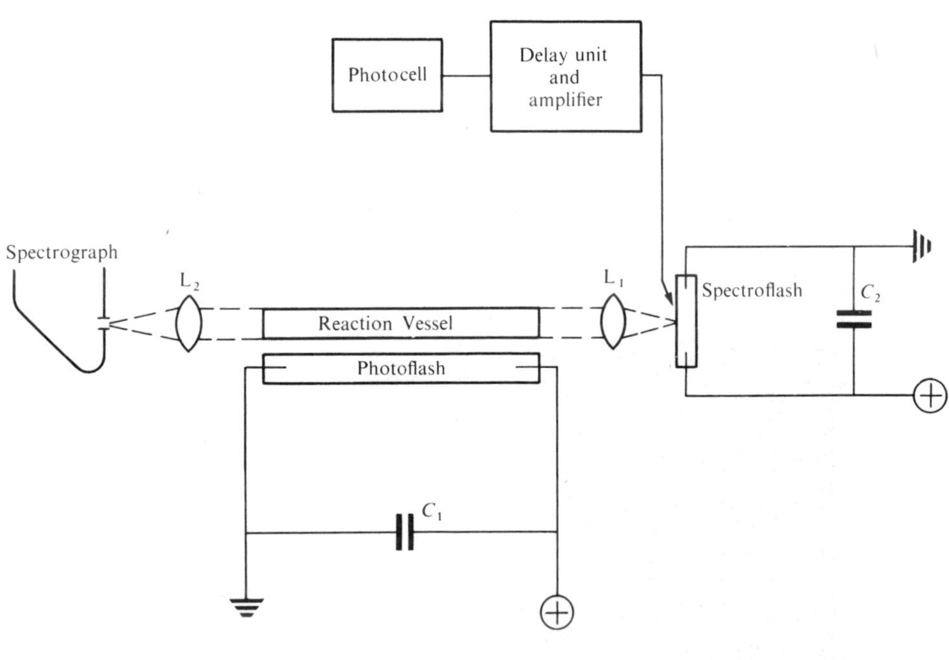

Figure 4.1
Schematic representation of a flash photolysis apparatus.

If transient species formed during the flash have suitable absorption spectra, the variation of spectral intensity with time can be followed. This intensity gives concentrations directly if the absorption coefficient is known. The method is known as *kinetic spectroscopy* and can be employed in one of two modes, called, by Porter, flash spectroscopy and kinetic spectrophotometry.

Flash spectroscopy

This is illustrated schematically in figure 4.1. The photoflash output is detected by a photocell and the resulting pulse is amplified to trigger a second flash lamp. The analysing flash or 'spectroflash' is similar to but smaller than the photoflash, with a discharge energy of around 100 J and a few microseconds duration. The spectroflash provides a continuum of background radiation against which the absorption spectra of species in the reaction vessel are photographed. This is achieved by placing the spectroflash at the focus of a lens (L_1) which gives an approximately parallel beam of background light. This passes through the reaction vessel and is then focused onto the slit of a spectrograph by a second lens (L_2). The photographic plate on the spectrograph records the absorption spectra at the time the spectroflash fired. This time is varied by the simple device of imposing a controlled and variable delay on the pulse that fires the spectroflash—the delay unit indicated in figure 4.1. The 'delay time' thus corresponds to the reaction time after the initiating photoflash and a series of experiments at different delay times provides a photographic record of transient species' spectra changing with time. A typical result is shown in figure 4.2.

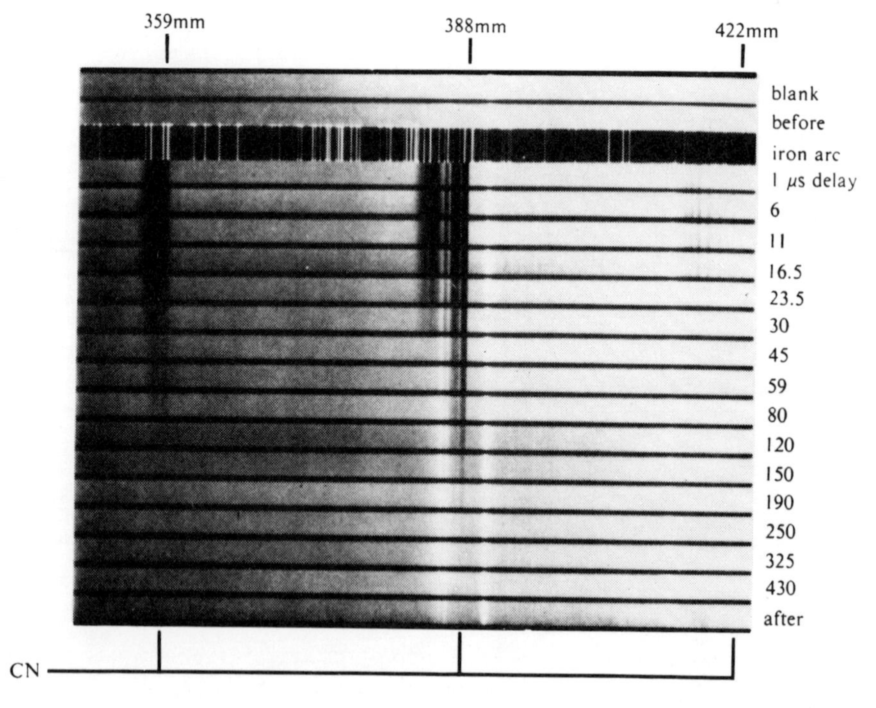

Figure 4.2
An example of flash spectroscopy. The formation and decay of the CN radical: $CNBr + h\nu \rightarrow CN^{\cdot} + Br^{\cdot}$; $2CN^{\cdot} + M \rightarrow C_2N_2 + M$. Spectra are taken first before the photolysis flash and then at the indicated delay times after the flash. Data from N. Basco, W. H. J. Vickers, J. E. Nicholas and R. G. W. Norrish, *Proc. R. Soc.*, **A272**, 147 (1963).

Kinetic spectrophotometry

In this mode, if the spectrum of the transient species is known, it is possible to obtain a complete kinetic run from one experiment. The spectroflash is replaced by a steady source of continuous background radiation extending over the appropriate wavelength range. The light passing through the reaction vessel is focused on to the slit of a monochromator fitted with a photoelectric detector set to the wavelength at which the transient absorbs, and the photocell output is fed to an oscilloscope. Thus the instrument is essentially a scanning spectrophotometer. The originally steady, background transmission at the pre-set wavelength falls when the concentration of absorbing species increases during the photoflash, but, as the transient species is removed in the subsequent reaction, the transmission increases until the steady background value is eventually reached. The oscilloscope trace thus records the formation and decay of the transient, as shown for a typical example in figure 4.3.

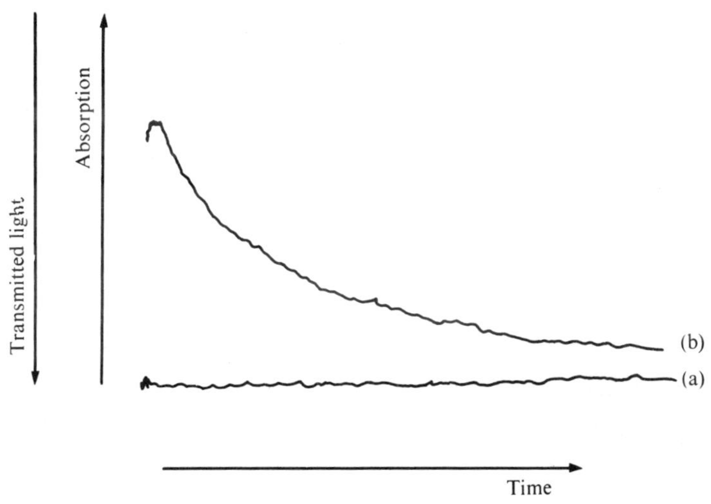

Figure 4.3
Kinetic spectrophotometry. Oscilloscope trace showing the decay of an absorbing species with time. (a) Represents the steady transmission when the photolysis flash is not fired; (b) shows absorption by the transient formed by the photolysis flash, and its subsequent decay.

The scope of the kinetic mode has recently been extended by the use of atomic resonance radiation, described in section 3.6. If appropriate atoms are formed in the flash, the corresponding atomic resonance radiation can be used as background source. The change in concentration of atoms in the reaction vessel is followed either from the change in absorption of this background radiation, as described above, or by detecting fluorescence from excited atoms in the reaction vessel.

Absorption of photoflash energy

The light energy absorbed by the system is rapidly degraded to heat by chemical reaction or collisional relaxation. If the vessel contents have very low heat capacity, as is the case for a gas at low pressure, the result is an almost instantaneous rise in temperature of several thousand degrees. Thus the system is subject to a considerable adiabatic shock, and these conditions may be used to study pyrolytic and explosive processes at high temperature. On the other hand, if a several hundredfold excess of inert gas is added to a low pressure of absorbing species to increase the total thermal capacity, the temperature rise can be kept below 10 K and reactions studied under reasonably isothermal conditions. These two general approaches are called the *adiabatic* and *isothermal methods* and they will be illustrated with a few examples from the enormous number available.

The adiabatic method

Among important reactions studied are the pyrolysis and combustion of hydrides[2]. When oxygen is added to hydrides to give combustion ignition is preceded by induction periods during which radical concentrations remain low, but on ignition radical concentrations, especially of hydroxyl, rise very rapidly.

Such studies are relevant to the burning of fuels, as in the internal combustion engine. With hydrogen or hydrocarbons, neither these 'fuel' molecules nor the 'oxidant' oxygen absorb radiation from the photoflash and a small concentration of photosensitizer must be added. This provides both adiabatic heating and radicals which initiate further reaction, and a good example is nitrogen dioxide ($NO_2 + h\nu \rightarrow NO + O^{\cdot\cdot}$). For hydrocarbons, flash photolysis clearly illustrates the effect of changing fuel: oxidant ratio. With oxygen-rich mixtures ignition is accompanied by a burst of hydroxyl radicals. With fuel-rich mixtures this is replaced by hydrocarbon radicals such as C_2, CH. 'Knock' and 'anti-knock' phenomena have also been investigated. Adding lead tetraethyl to mixtures such as those described above lengthened the induction periods, reflecting the decreased tendency to premature ignition or knocking. The spectrum of gaseous lead oxide was seen during the induction period, but on ignition this was replaced by atomic lead. Such evidence shows that premature ignition is reduced by reactions of lead with hydroxyl and hydrocarbon radicals (see section 7.10).

The isothermal method

Examples will be chosen from radical reactions and later from energy transfer studies, and the radical systems selected are relevant to other sections in the book.

(a) Halogen molecules absorb light in the u.v. or visible and dissociate, the intensity of the molecular spectrum being reduced as the concentration falls. Halogen atoms formed in the flash re-combine over a period of a few milliseconds, and the process is easily followed by observing the resultant increase in intensity of the halogen molecular spectrum:

$$X_2 + h\nu \rightarrow 2X^{\cdot}$$

$$X^{\cdot} + X^{\cdot} + M \rightarrow X_2 + M$$

Thus, the combination rate constant is easily measured and the activation energy can also be determined by enclosing the system in a suitable furnace to vary the temperature. This gave the strange result that as temperature increased the rate decreased—the reaction has a *negative* activation energy. The rate constant has been measured at much higher temperatures using shock tubes (section 4.5) with consistent results. The phenomenon is discussed in section 5.8. The third-order rate constant can be measured

[2] For a summary, see R. G. W. Norrish, *Chemistry in Britain*, **1**, 289 (1965).

for many different third bodies (M) thus giving their relative efficiencies which are found to vary widely[3].

(b) With exothermic reactions of the type:

$$A + BCD \rightarrow AB + CD$$

the molecule with the newly formed bond takes a high proportion of the exothermicity as vibrational energy, e.g. in the photolysis of nitrogen dioxide the spectrum of vibrationally excited oxygen molecules is observed[4]:

$$NO_2 + h\nu \rightarrow NO + O^{\cdot\cdot}$$

$$O^{\cdot\cdot} + NO_2 \rightarrow NO + O_2^{\dagger}$$

(c) Detection by atomic resonance fluorescence has been used to study an ozone-forming reaction with very low concentrations of oxygen atoms[5]:

$$O^{\cdot\cdot} + O_2 + M \rightarrow O_3 + M$$

This reaction is important in maintaining the stratospheric ozone balance, about which there is serious concern at present (see section 6.2).

Energy transfer studies can be illustrated with examples chosen for the decay of: (a) vibrationally excited states; and (b) metastable electronic states.

(a) Nitric oxide absorbs u.v. radiation and is excited to an upper electronic state from which fluorescent decay to the ground state is rapid ($\sim 10^{-8}$ s). Studying fluorescence shows that this gives ground state molecules in vibrational levels from $v = 0$ to $v = 5$, as discussed in section 2.8. During the microsecond flash excitation and decay occur many times and give considerable overpopulations of these vibrationally excited states. However, flash photolysis finds a strong absorption spectrum only for the first excited vibrational state, which persists for about 200 μs.[6] This indicates that higher vibrational states decay by rapid 'near-resonance' energy transfer:

$$NO(v = n) + NO(v = n - 2) \rightarrow 2NO(v = n - 1)$$

Such efficient steps halt at $NO(v = 1)$ which must decay by vibrational–translational transfer

$$NO(v = 1) + M \rightarrow NO(v = 0) + M$$

The rate constants for this process have been calculated with a series of second bodies (M). The probability of transfer per collision varies from $3{\cdot}6 \times 10^{-4}$ for $M = NO$ to 4×10^{-7} for $M = N_2$.

(b) The collisional transfer of electronic energy has recently been studied for such atoms as Pb, P and I using detection by atomic resonance radiation[7].

With most electronically excited molecules, decay is too rapid for the time resolution of this method. However, metastable triplet states (see section 2.9) of polyatomic

[3] For example, G. Porter and J. A. Smith, *Proc. R. Soc. (London)*, **A261**, 28 (1961).
[4] F. J. Lipscomb, R. G. W. Norrish and B. A. Thrush, *Proc. R. Soc. (London)*, **A233**, 455 (1956).
[5] R. E. Huie, J. T. Herron and D. D. Davies, *J. Phys. Chem.*, **76**, 2653 (1972).
[6] N. Basco, A. B. Callear and R. G. W. Norrish, *Proc. R. Soc. (London)*, **A260**, 459 (1960).
[7] D. Husain and J. G. F. Littler, *J. Chem. Soc., Faraday Trans.* II, **68**, 2110 (1972).

molecules have been studied extensively in solution, and observed for such molecules as naphthalene and anthracene, in the gas phase. For detailed gas phase studies, nanosecond pulsed lasers provide a more sensitive probe.

4.2 Pulsed Lasers (Nanosecond)

With advances in laser technology it became possible to extend the range of flash photolysis from the microsecond to the nanosecond region[8], and ultimately to picosecond kinetics.

Solid state lasers such as ruby or neodymium glass can be modified to give a light pulse lasting only a few nanoseconds. The laser is excited by a flash lamp of the type described previously, but a Q-switch (such as a bleachable dye) placed in the light path prevents lasing action until the dye is bleached and starts to transmit light. By this time there is a considerable overpopulation of excited energy states in the lasing material and stimulated emission builds up rapidly giving very high amplification of emitted light. This results in a giant laser pulse of several joules energy lasting less than 20 ns. The emitted radiation is monochromatic, 694 nm with ruby and 1·06 μm with neodymium glass. These wavelengths are too high (radiation too low in energy) to initiate photochemically interesting processes, but if the pulse is passed through certain types of crystal, e.g. potassium dihydrogen phosphate for ruby, the phenomenon of frequency doubling occurs, and up to 20% of emerging radiation is the first harmonic, 347 nm for ruby. With neodymium glass, frequency doubling twice in succession gives about 1% of the initial radiation in a pulse at 265 nm and further amplification is obtained by passing through a second laser rod. The experimenter's next problem is to follow changes in transient concentration over the nanosecond time scale, and methods analogous to those described above for flash photolysis have been developed.

Flash spectroscopy

A very short-lived analysing background flash is required to build up a series of pictures of the transient's absorption spectrum at different times after the initiating pulse. Such a flash can be produced by using a beam splitter which transmits part of the pulse directly to the reaction vessel and also reflects part of the initiating pulse into a scintillation solution as shown in figure 4.4. The solution is highly excited and fluoresces rapidly, emitting a continuum at longer wavelengths than that of the exciting pulse. This induced flash is of the same duration, several nanoseconds, as the initiating pulse and so is a suitable background analysing flash or spectroflash. It passes through the reaction vessel to a spectrograph. The time delay between irradiation of the reaction vessel by the initiating pulse and the arrival of the spectroflash is controlled by the velocity of light, thus eliminating electronics from the analysis system. The delay time is simply the time taken for the pulse to travel to the scintillator solution added to the time taken for the spectroflash to travel to the reaction vessel. The total light path is easily varied to give time delays in the nanosecond range (3 m corresponds to 10 ns delay).

[8] G. Porter and L. Patterson, *Chemistry in Britain*, **6**, 246 (1970).

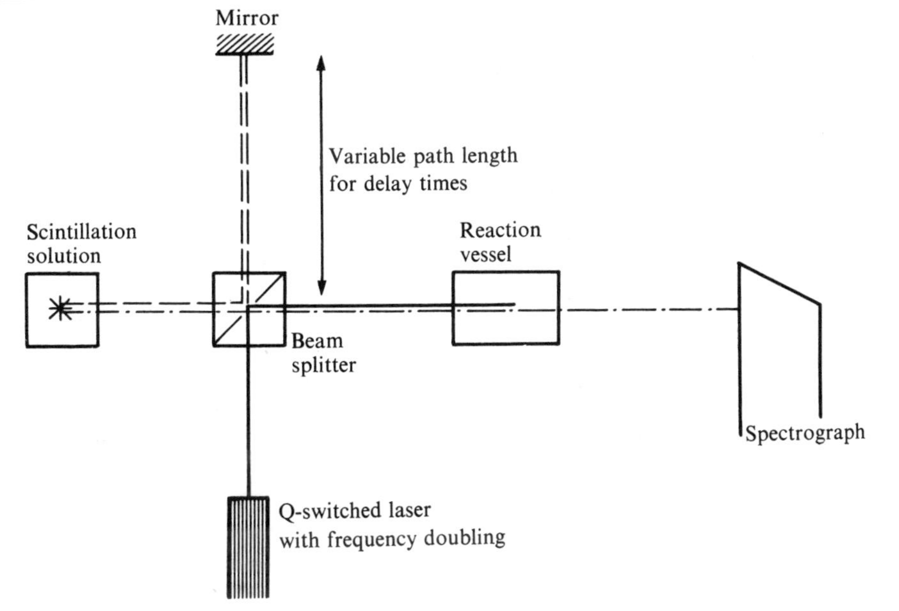

Figure 4.4
Schematic representation of apparatus for flash spectroscopy with nanosecond pulsed laser.
——— path of initiating pulse;
- - - - - - path of pulse from beam splitter to scintillating solution;
- — - — - path of spectroflash pulse.

Kinetic spectrophotometry

Transient concentrations are followed by providing steady background light and continuously recording the change in light intensity at the wavelength where the transient absorbs. A conventional flash discharge of several hundred microseconds duration is sufficiently steady during the period of kinetic interest—several nanoseconds to a few microseconds—and gives background light of suitably high intensity. The 'steady' discharge light can be fired synchronously with the laser pulse.

A pulsed laser is monochromatic and this helps provide a detailed insight into the energetics of processes studied. However, the range of available laser frequencies represents a limited choice for the photochemist, although it is being extended. Kinetic spectroscopy can be applied to new types of process in the nanosecond time range. A prominent example is the decay of excited *singlet* states in condensed phases, with lifetimes such as 45 ns for triphenylene[9]. As described in section 2.9, molecules can undergo radiationless transfer (intersystem crossing) from excited singlet states to high vibrational levels of a triplet state. This is followed by decay to the lowest vibrational level of the triplet in the rapid radiationless process. This vibrational relaxation has been studied for the triplet state of anthracene in the gas phase[10].

[9] G. Porter and M. R. Topp, *Proc. R. Soc.* (*London*), **A315**, 163 (1970).
[10] S. J. Formosinho, G. Porter and M. A. West, *Chem. Phys. Lett.*, **6**, 7 (1970).

4.3 Pulsed Lasers (Picosecond)

A sophisticated variant of the pulsed laser technique has now successfully probed the picosecond time scale, observing relaxation processes in electronically excited states including radiationless transitions. The technique called 'mode locking' gives, instead of a single nanosecond pulse, a train of equally spaced picosecond pulses, each of high power[11].

The effect of two successive pulses on an absorbing molecule has been used ingeniously to study the rate of transitions between excited electronic states. In some molecules absorption of a photon from the first pulse gives the lowest excited singlet state and absorption of a second photon from a following pulse gives a higher singlet. The molecule decays from the upper state by fluorescence, and from the way fluorescent intensity varies with the interval between pulses the kinetics of intersystem crossing from the lower singlet can be determined in the picosecond range.

Experimenters have now succeeded in isolating one of the train of picosecond pulses[12]. This allows use of flash spectroscopy in a manner similar in principle to that described above for nanosecond pulses. The isolated pulse passes through a beam splitter, one pulse going directly to the reaction vessel, the other being reflected to a cell containing water or D_2O. The resulting flash of radiation is continuous in the range 300–600 nm, with a duration of a few picoseconds. This is used as analysing flash and is reflected to the reaction vessel after a delay time controlled by the optical path length and the velocity of light. The technique is being applied to study energy transfer within single long chain molecules such as polymers and biological molecules, and to a variety of fast processes in solution.

4.4 Pulse Radiolysis

This is the radiation chemistry analogue of flash photolysis, initiation being provided by a pulse of ionizing radiation such as high energy electrons. The duration of such pulses has improved from microseconds to nanoseconds and picoseconds, and transient species are observed in the spectroscopic or kinetic modes as described above for flash photolysis. The method has found its greatest application in radiation chemistry which lies outside the scope of this book[13], but there have been applications to radical reactions in the gas phase, for example studies of NH_2 from ammonia and CH from methane or acetylene.

4.5 Shock Tubes

This is a technique for studying reactions in the millisecond–microsecond range at high temperature[14]. The reactants are heated by a shock wave travelling at supersonic

[11] P. M. Rentzepis, *Science*, **169**, 239 (1970).
[12] P. M. Rentzepis *et al.*, *Ann. Rev. Phys. Chem.*, **24**, 473 (1973).
[13] A recent review is L. M. Dorfman, *Techniques of Chemistry*, Vol. 6, Part 2, 3rd ed. (Ed. E. S. Lewis), Wiley Interscience, 1974.
[14] J. N. Bradley, *Chemical Applications of the Shock Tube*, *Royal Inst. Chem. (London) Lectures*, Series No. 6, **1** (1963).

speed, the heating taking place in a time that corresponds to a few molecular collisions. The high temperature stays approximately constant for hundreds of microseconds or a few milliseconds in some cases, and throughout this period the concentration of radicals is followed by one of the standard methods of detection.

The essential features of the apparatus are shown in figure 4.5. The shock tube is a metal or glass pipe several centimetres in diameter and several metres long. It is divided into two sections by a thin metal or plastic diaphragm. One section contains the 'driver' gas, usually hydrogen or helium at pressures up to ten times atmospheric pressure. The other section contains a reactant gas at a few torr, a few hundred N m^{-2} pressure. Towards the end of the tube is the observation point for measurement of radical concentrations, and detectors for measuring the velocity of the shock front.

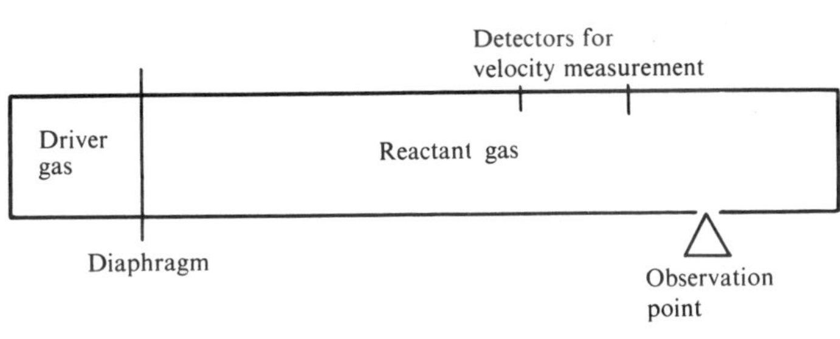

Figure 4.5
Schematic representation of shock tube.

When the diaphragm is mechanically ruptured the driver gas bursts out and rapidly forms a shock wave with a self-sustaining, sharp shock front. The front travels with supersonic speed compressing and heating the reactant gas as it proceeds. The variation of pressure along the tube is represented in figure 4.6. This shows the regions: A, unexpanded driver gas; B, expanded driver gas; C, the already shocked reactant, extending from the shock front to the surface between driver gas and reactant; D, reactant gas as yet unshocked. It is in C that the high-temperature reactions take place, and as the shock front advances this reaction zone is pushed down the tube.

Eventually the shock front reaches the observation point. This has previously been recording zero radical concentration or steady reactant concentration, but as the front arrives the reactant gas at the observation point is suddenly compressed and heated. This corresponds to the initial time for the observed reaction. The reaction zone is then pushed past the observation point—the gas observed is that which was shocked earlier and so has been reacting at the high temperature for longer. Thus the recorded concentrations correspond to progressively longer reaction times until eventually the driver gas arrives. The last element of reaction zone observed is that which was shocked first, as the diaphragm burst. The time between the arrival of the shock front and the arrival of the driver gas is the maximum reaction time that can be studied. It is usually about a millisecond, whereas the 'initiation time' for heating an element of the reaction gas is extremely short (the process requires only a few molecular collisions).

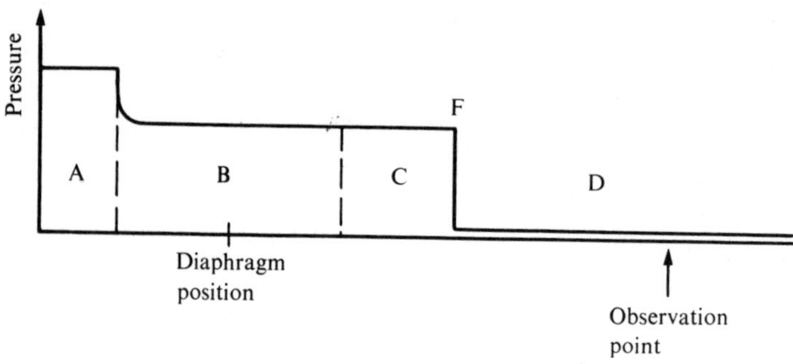

Figure 4.6
The variation of pressure along the shock tube after rupturing of diaphragm. A, unexpanded driver gas; B, expanded driver gas; C, shocked reactant gas; D, unshocked reactant gas; F, shock front.

The diagram is not drawn to scale, reaction zone C in particular being greatly exaggerated in width.

The temperature in the reaction zone is calculated with an accuracy of ± 50 K from the initial conditions in the reactant gas and the velocity of the shock wave. The latter is measured during the experiment by recording the time taken for the front to pass two detectors in the tube wall—the detectors may simply be sensitive, temperature responsive probes or pressure sensitive, piezoelectric crystals. Typical temperatures for the reaction zone are of the order of a few thousand degrees, but up to 20 000 K is accessible.

Various methods may be employed to follow changes in concentration and the most popular are optical spectroscopy and mass spectrometry. Many species are in excited states at the high temperature, and emission spectra may be observed through a suitable window. Alternatively, absorption spectra may be observed with a background light source behind a second window. For mass spectrometry a 'pin-hole' leak direct to the spectrometer is usual.

The shock tube may be used in other modes:

(i) When the shock front reaches the end of the tube it is reflected back through the oncoming compressed gas, subjecting it to further shock and heating, and the apparatus can be arranged to study reactions following the reflected shock. This usually requires a system where little reaction occurs in the first shock, substantial reaction only taking place at the higher temperature (approximately doubled) behind the reflected shock.

(ii) The single-pulse method. As the first shock wave travels down the tube, a rarefaction wave moves in the opposite direction. This is reflected off the back wall and then travels in the same direction as the shock wave. In a suitably modified apparatus the rarefaction wave eventually interacts with the shock and quenches the reacting mixture after a calculated reaction time. The products can then be analysed, say by g.l.c. or mass spectrometry, as in a *static* system. The experiments can be repeated at a series of reaction times obtained by varying the driver and reactant pressures.

A few illustrative examples of the types of reaction investigated with shock tubes are:

(i) elementary radical reactions such as[15]

$$O^{\cdot\cdot} + CO + Ar \rightarrow CO_2 + Ar$$

(ii) dissociation reactions such as[16]

$$HCl + M \rightarrow H^{\cdot} + Cl^{\cdot} + M$$

(iii) complex systems can be probed, for example the hydrogen–oxygen reaction can be followed by observing the absorption spectrum of the hydroxyl radical. This type of investigation supplements the contribution of other techniques, such as flash photolysis or flame studies, to understanding the results of the enormous amount of work carried out in static systems (see chapter 7).

(iv) Energy transfer processes, particularly vibrational–translational transfer can be investigated in the region immediately following the shock front. When an element of gas is shocked, translational–translational and translational–rotational transfer are so efficient that equilibrium at the high temperature is reached in less than ten molecular collisions. Vibrational– translational transfer is much less efficient, a typical probability of transfer per collision being 10^{-4}. The build-up of vibrationally excited states to their high-temperature equilibrium concentration occurs over a period of many microseconds and may be followed by absorption spectroscopy.

4.6 Fast-Flow Methods

Fast-flow resembles conventional flow methods described in section 3.13, but it uses a rapid flow of reactants at very low total pressure together with special mixing devices that restrict mixing times to under one millisecond. This means that elementary radical reactions with half-lives of a few milliseconds and above can be studied.

The apparatus is represented schematically in figure 4.7. In essence it consists of some method for generating atoms or simple radicals which are then mixed with a separate flow of the stable reactant. The mixture flows down a glass tube, perhaps 2 cm in diameter and 1 m long, past some detection device. Gas flow is controlled by needle valves and the flow rate, usually a few hundred cm s^{-1}, is measured by a flow meter. Gas pressures in the flow tube are in the range 0·1–10 Torr, 13–1·3 × 10^3 N m^{-2}, and specially constructed diffusers at the point of entry ensure efficient mixing. The reaction starts on mixing and the concentration of reactants or products is determined at the observation point downstream. Under steady-flow conditions this corresponds to the reaction time $t = d/v$ where d is the distance between mixing point and detector, and v is the linear flow velocity. The reaction time is varied by altering d, for example by placing the reactant entry tube down the central axis of the flow tube, the reactant tube being movable along the axis. This is the type of arrangement shown in figure 4.7. Alternatively there may be a series of fixed reactant entry points at various positions along the flow tube. To determine activation energies the flow tube is enclosed within a thermostatted bath or furnace.

[15] M. C. Lin and S. H. Banes, *J. Chem. Phys.*, **50**, 3377 (1969).
[16] D. J. Seery and C. T. Bowman, *J. Chem. Phys.*, **48**, 4314 (1968).

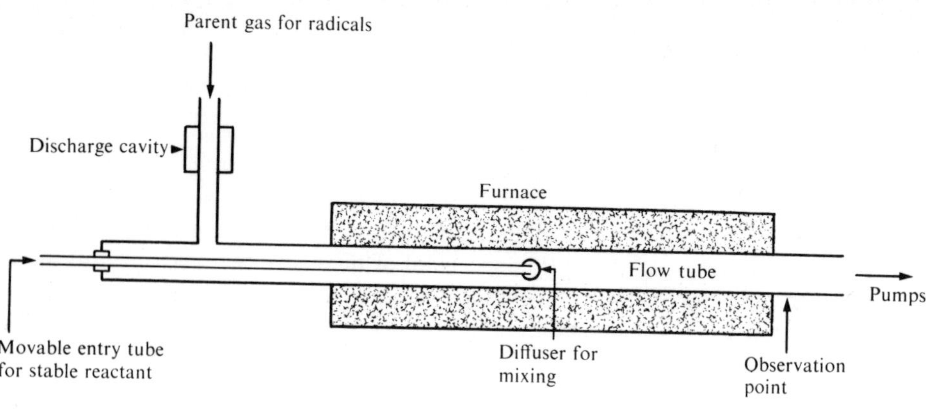

Figure 4.7
Schematic representation of fast-flow system.

Radicals are usually produced by flowing the parent gas through a microwave electrical discharge. Atoms such as hydrogen, oxygen, nitrogen and halogens are formed in high relative concentrations from the diatomic molecules, sometimes with inert gas present as carrier. Great care is necessary with very low pressure to prevent the removal of atoms at the wall of the flow tube and treatment of surfaces with phosphoric or boric acid has proved successful in this respect.

Many of the detection methods discussed in sections 3.5–10 have been employed in fast-flow systems[17]. They include:

(i) absorption spectra, e.g. for the reactions

$$CN^{\cdot} + O_2 \rightarrow OCN^{\cdot} + O^{\cdot\cdot}$$

$$ClO^{\cdot} + H_2 \rightarrow HOCl + H^{\cdot}$$

(ii) mass spectrometry, e.g.

$$OH^{\cdot} + OH^{\cdot} \rightarrow H_2O + O^{\cdot\cdot}$$

$$H^{\cdot} + CH_3CHO \rightarrow CH_4 + CHO^{\cdot}$$

(iii) calorimetric probes, e.g.

$$H^{\cdot} + H^{\cdot} + H_2 \rightarrow H_2 + H_2$$

However, the methods probably most associated with fast flow are e.s.r., chemical titration, chemiluminescence, and most recently atomic resonance fluorescence.

[17] References to original work on these and other reactions are found in M. A. A. Clyne, *Chem. Soc. Ann. Rep.* (A), **65**, 167 (1968), and R. J. Donovan, D. Husain and L. J. Kusch, *Chem. Soc. Ann. Rep.* (A), **69**, 19 (1972).

(iv) E.s.r. Here the flow tube passes through an e.s.r. cavity and among the many reactions for which it has been used are:

$$O^{\cdot\cdot} + H_2 \rightarrow OH^{\cdot} + H^{\cdot}$$

$$H^{\cdot} + C_2H_4 \rightarrow C_2H_5^{\cdot}$$

$$H^{\cdot} + HCl \rightarrow H_2 + Cl^{\cdot}$$

$$H^{\cdot} + O_2 + M \rightarrow HO_2 + M$$

(v) Chemical titration. Here the atom whose concentration is to be determined must react faster with the titrant gas than with any other species present. The titrant is added just upstream of the detector and its flow gradually increased until the end-point is reached. The critical titrant concentration is then equivalent to the atom concentration originally present. Suitable titrants and the technique for detecting the end-point depend on the particular reaction investigated. A classic example is the use of nitrogen dioxide as titrant to determine oxygen atom concentration, as described in section 3.7. Nitrogen atoms may be titrated with nitric oxide:

$$N + NO \rightarrow N_2 + O$$

$$N + O \rightarrow NO^*$$

The blue glow emitted from excited nitric oxide molecules is extinguished when no nitrogen atoms remain. Some reactions that have been followed by the titration method are:

$$O + O + M \rightarrow O_2 + M$$

$$N + N + M \rightarrow N_2 + M$$

$$N + O_2 \rightarrow NO + O$$

(vi) Chemiluminescence. Hydrogen atom concentration can be determined by adding nitric oxide just upstream of a photoelectric detector. HNO is formed in an electronically excited state:

$$H^{\cdot} + NO \rightarrow HNO^*$$

and emits a red glow whose intensity is measured photoelectrically. As opposed to titrating to an end-point, here the concentration of atoms is calculated from the intensity of the chemiluminescence. The intensity is proportional to concentration: $I = K[NO][H]$ and K is established by independent calibration. An example of a reaction monitored in this way is

$$H^{\cdot} + O_2 + M \rightarrow HO_2^{\cdot} + M$$

Another example where chemiluminescent detection has been employed is

$$S^{\cdot\cdot} + S^{\cdot\cdot} + M \rightarrow S_2^* + M$$

(vii) Atomic resonance fluorescence. The method described in section 3.6 is providing a valuable extension to the fast-flow method. At the relatively high radical concentrations required for detection by other means, the half-lives of many second order radical reactions are in the microsecond range, too short for the time resolution of the fast-flow method. Due to the extra sensitivity of atomic resonance fluorescence, much lower radical concentrations can be employed and these half-lives, which are inversely proportional to concentration, can come into the millisecond range and so within the scope of fast flow. Reactions so studied include:

$$Cl^{\cdot} + ClNO \rightarrow NO + Cl_2$$

$$O^{\cdot\cdot} + NO_2 \rightarrow NO + O_2$$

$$Br^{\cdot} + IBr \rightarrow Br_2 + I^{\cdot}$$

Further Reading

Comprehensive Chemical Kinetics, Vol. 1, (Eds C. H. Bamford and C. F. H. Tipper), Elsevier, 1969.

Nobel Symposium 5, Fast Reactions and Primary Processes in Reaction Kinetics, (Ed. S. Claesson), Interscience, 1967 (articles by R. G. W. Norrish and G. Porter on flash photolysis).

G. Porter and M. A. West, *Techniques of Chemistry*, Vol. 6, Part 2, 3rd ed., (Ed. G. G. Hammes), Wiley Interscience, 1974 (flash photolysis, microsecond, nanosecond and picosecond pulsed lasers).

P. M. Rentzepis *et al.*, *Ann. Rev. Phys. Chem.*, **24**, 473 (1973) (picosecond flash photolysis).

E. F. Greene and J. P. Toennies, *Chemical Reactions in Shock Waves*, Academic Press, 1964.

J. N. Bradley; *Fast Reactions*, Oxford University Press, 1975.

Exercises

4.1
In a flash photolysis experiment a radical, $R^{\cdot}$, was produced during the flash (of duration 2 μs) and its subsequent decay, $2R^{\cdot} \rightarrow R_2$, was followed by observing the decay in absorbance of the radical spectrum. The results are given in the table:

Time/μs	0	10	15	20	25	30	40	50	75	100
Absorbance	0·75	0·58	0·51	0·47	0·41	0·38	0·32	0·28	0·20	0·17

The path length of the photolysis vessel was 50 cm and the absorption coefficient for the radical is $1{\cdot}1 \times 10^4$ mol^{-1} 1 cm^{-1}. Evaluate the rate constant for radical combination.

(Ans: $2{\cdot}6 \times 10^{13}$ mol^{-1} cm^3 s^{-1})

4.2
In the flash photolysis of iodine–argon mixtures the formation and decay of iodine atoms was monitored by kinetic spectrophotometry of I_2 absorption. In each experiment the argon was present in large excess and the rate of removal of iodine atoms was found to be second order in iodine atom concentration. Repeating the experiment with different argon pressures showed that the reaction was first order with respect to argon. The experimental third-order rate constant, however, varied with the iodine: argon ratio as shown in the table:

$([I_2]/[Ar])/10^{-3}$	3·36	2·25	1·64	1·18	0·52	0·22
$k_{exp}/10^{15}$ mol^{-2} cm^6 s^{-1}	12·7	10·4	8·91	8·68	7·24	6·77

Evaluate the rate constants for iodine atom combination: (a) with argon as third body; (b) with iodine as third body.

(Ans: (a) $6{\cdot}4 \times 10^{15}$ mol^{-2} cm^{6} s^{-1}; (b) $1{\cdot}8 \times 10^{18}$ mol^{-2} cm^{6} s^{-1})

4.3

The single-pulse shock-tube method was used to study the unimolecular decomposition $C_2H_5I \xrightarrow{k_a} C_2H_4 + HI$. To avoid uncertainties in calculating temperature a comparative study was carried out for $C_3H_7I \xrightarrow{k_b} C_3H_6 + HI$ for which it was known that $k_b = 9{\cdot}1 \times 10^{12}\ e^{-E_b/RT}$ s^{-1}, where $E_b = 182$ kJ mol^{-1}. For a reaction time of 0·22 ms, 5·0% decomposition occurred. What was the temperature in the shock wave? Under these conditions the decomposition of C_2H_5I was 0·90%. Evaluate k_a at this temperature. At 800 K the ratio of rate constants was $k_a/k_b = 0{\cdot}102$. What are the Arrhenius parameters for k_a?

(Ans: 900 K; $A_a = 5{\cdot}8 \times 10^{13}$ s^{-1}; $E_a = 210$ kJ mol^{-1})

4.4

When hydrogen is added to oxygen atoms, previously generated in a fast-flow system, the relatively slow reaction, $O + H_2 \xrightarrow{k} OH + H$ takes place. This is followed by the very fast reaction, $OH + O \rightarrow O_2 + H$. The concentration of oxygen atoms can be followed by observing the intensity (I) of light emitted when nitric oxide is added to the flowing gas, and $I \propto [O][NO]$.

With hydrogen present in excess, plots of $\ln\{I/[NO]\}$ versus t gave straight line graphs of the following slopes:

$[H_2]/10^{-2}$ Torr	2·45	3·52	4·60	6·05	8·03
Slope/s^{-1}	3·84	5·63	7·38	9·38	11·36

Evaluate k.

(Ans: $1{\cdot}3 \times 10^{9}$ mol^{-1} cm^{3} s^{-1})

4.5

The equilibrium constant for the dissociation of diatomic molecule X_2:

$$X_2 + X_2 \underset{k_2}{\overset{k_1}{\rightleftharpoons}} 2X^{\cdot} + X_2$$

is $K = 1{\cdot}0 \times e^{-E/RT}$ mol cm^{-3} where $E = 180$ kJ mol^{-1}. X_2 exhibits strong continuous absorption in the visible region of the spectrum. The equilibrium constant for dissociation of diatomic molecule Y_2 is $K' = 10 \times e^{-E'/RT}$ mol cm^{-3}, where $E' = 450$ kJ mol^{-1}. The predominant reaction of X atoms with Y_2 is thought to be $2X^{\cdot} + Y_2 \xrightarrow{k_3} 2XY^{\cdot}$ where $k_3 \gg k_2$. The subsequent reaction of radical XY is postulated as $2XY^{\cdot} \xrightarrow{k_4} X_2 + Y_2$.

How would you employ the techniques described in this chapter to study X_2 alone and X_2–Y_2 mixtures to confirm the mechanism suggested above and measure rate constants k_1 to k_4: (a) at 300 K; (b) at 1500 K?

Chapter 5: Theories of Reaction Rates

Employing the range of experimental techniques described in the previous chapters, rate constants and activation energies have been measured for a tremendous number of elementary reactions in the gas phase. Virtually all the constants conform to the Arrhenius equation, $k = A_{\text{exp}}\, \text{e}^{-E_{\text{exp}}/RT}$ and the more important attempts to predict values of A_{exp}, E_{exp} and so k, from fundamental molecular properties are discussed in this chapter (with the exception of Molecular Dynamics Theory described in chapter 8). Results for the pre-exponential factor, A_{exp}, have often been quite satisfactory, but much empiricism is still involved in interpreting the activation energy term.

The simplest approach to calculating rate constants for bimolecular reactions associates A with the frequency of collisions between reactant molecules and E with an energy barrier to reaction, probing no further in this direction. This is the Simple Collision Theory, the earliest rate theory to achieve significant success.

Other rate theories start with a more detailed account of how the system's potential energy changes as reactants are converted to products. It is convenient to map out the relationship between potential energy and interatomic distances in the form of a potential energy surface, on which an energy barrier to reaction usually appears. The first important alternative to the Simple Collision Theory considered the most probable route from reactants to products across the energy surface passing over the energy barrier. The molecular configuration at the top of the barrier is called the activated complex (or transition state, see section 5.5) and, in calculating rate constants by Activated Complex Theory, a quasi-equilibrium between reactants and the activated complex is assumed and equilibrium statistical mechanics applied.

This theory dominated the field for thirty years, but recent advances in rapid computational methods have made a more sophisticated approach possible. Molecular

Dynamics Theory calculates the probability of reaction on passing over the energy surface for a variety of initial reactant conditions, and rate constants are computed from appropriate average initial conditions. This chapter is concerned with the earlier theories which still have the widest applicability in kinetics. Most elementary reactions are bimolecular and are free from some extra complications found with unimolecular and termolecular processes, so the theoretical discussion will start with the bimolecular case.

5.1 Simple Collision Theory for Bimolecular Reactions

In the Arrhenius formulation the rate constant combines an exponential dependence on an energy term and a pre-exponential factor, $k = A_{\text{exp}}\, e^{-E_{\text{exp}}/RT}$. In the late nineteenth century both Arrhenius and van't Hoff suggested the energy term represented an energy barrier to reaction and this was incorporated into the theory when, in the second decade of this century, both Trautz and Lewis[1] treated the pre-exponential factor as the rate at which molecules collide. This frequency of collision (pairs of molecules per unit volume per second) is calculated easily if the molecules are regarded as simple hard spheres. The basic model for the Simple Collision Theory (SCT) may be expressed, for the reaction $A + B \rightarrow$ products, as

$$\begin{aligned} \text{Reaction rate} &= \text{Collision frequency} \times \begin{pmatrix} \text{fraction of collisions} \\ \text{with sufficient energy} \\ \text{for reaction} \end{pmatrix} \\ &= \qquad Z_{AB} \qquad \times \qquad F \end{aligned} \tag{5.1.1}$$

The two terms on the right-hand side of (5.1.1) will now be calculated separately.

(i) Collision frequency, Z_{AB}

Rigorous derivations appear in standard texts on the kinetic theory of gases[2] and the kineticists' favourite approximate method is given here. This is based on considering average collision properties, which, due to the extremely high numbers of collisions per second, are very well defined. Due to the convenience of cm^3 as volume unit in gas kinetics, velocity will be expressed as cm s^{-1}. If the average relative velocity for A and B molecules is $\bar{v}_{AB}$ cm s^{-1}, an *average* situation for a typical A molecule is where it travels at $\bar{v}_{AB}$ whilst all B molecules in its vicinity are stationary. This is represented in figure 5.1, where the path of A is indicated, and an imaginary cylinder has been constructed with this path as its axis. The cylinder's radius is the distance between the centres of an A molecule and a B molecule when they just touch. The molecules are regarded as simple hard spheres of radii r_A and r_B cm. The cross sectional area of the cylinder is thus $\pi(r_A + r_B)^2 = \sigma_{AB}$. The term σ_{AB} is called the cross sectional area for collision or collisional cross section. If the distance of closest approach (centre-to-centre) is greater than $(r_A + r_B)$, hard spheres do not collide. The length of the cylinder is $\bar{v}_{AB}$ cm and so its volume is $\sigma_{AB}\bar{v}_{AB}$ cm^3.

In one second this typical A travels a distance $\bar{v}_{AB}$ and will collide with all B molecules whose centres lie within the cylinder. Thus the frequency of collisions, Z_A, between the

[1] M. Trautz, *Z. Anorg. Chem.*, **96**, 1 (1916); W. C. McC. Lewis, *J. Chem. Soc.*, **113**, 471 (1918).
[2] E. A. Moelwyn-Hughes, *Physical Chemistry*, Pergamon, Oxford, 1957.

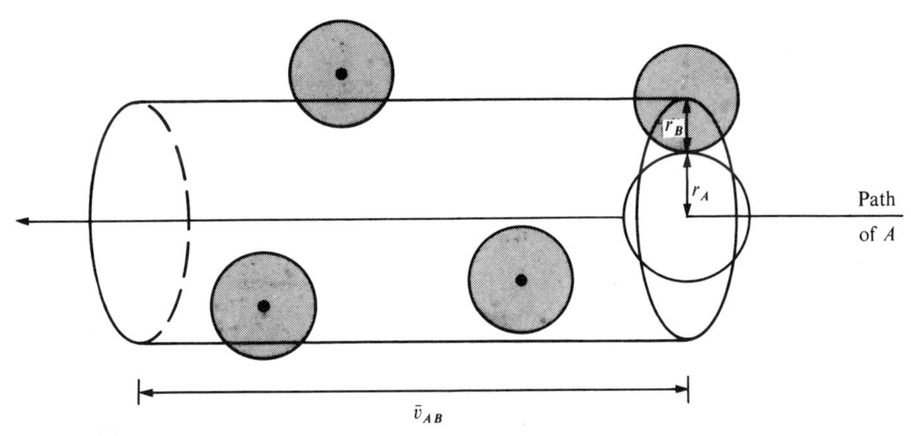

Figure 5.1
Diagram of model for calculating frequency of collisions. A cylinder length $\bar{v}_{AB}$ and radius $(r_A + r_B)$ is drawn with the path of A as its axis.

single A molecule and B molecules is given by the volume of the cylinder multiplied by the number of B molecules per cm^3, C_B:

$$Z_A = \sigma_{AB}\bar{v}_{AB}C_B \text{ s}^{-1}$$

The model is artificial in that A should change direction on colliding with B, but it yields the correct average result.

The total frequency of bimolecular collisions between all A and B molecules, Z_{AB}, is the collision frequency for the single, typical A multiplied by the number of A per cm^3, C_A:

$$Z_{AB} = \sigma_{AB}\bar{v}_{AB}C_AC_B \text{ molecules cm}^{-3}\text{ s}^{-1} \tag{5.1.2}$$

or

$$Z_{AB} = Z_{AB}{}^0C_AC_B \tag{5.1.3}$$

where $Z_{AB}{}^0$ is called the *collision number*, the collision frequency at unit concentrations. In a pure gas, A, the frequency of collisions, Z_{AA}, is:

$$Z_{AA} = \tfrac{1}{2}\sigma_{AA}\bar{v}_{AA}C_A{}^2 \text{ molecules cm}^{-3}\text{ s}^{-1} \tag{5.1.4}$$

The factor, $\frac{1}{2}$, occurs because where A and B are identical the above treatment counts each collision twice.

Average relative velocities are easily deduced from the Maxwell–Boltzmann velocity distribution (section 2.3) as a function of universal constants and molecular masses alone. If μ is the reduced mass, defined by $1/\mu = (1/m_A) + (1/m_B)$ where m_A, m_B are the molecular masses, then:

$$\bar{v}_{AB} = \left(\frac{8\bar{k}T}{\pi\mu}\right)^{1/2} \qquad \bar{v}_{AA} = \left(\frac{16\bar{k}T}{\pi m_A}\right)^{1/2} \tag{5.1.5}$$

When calculating these velocities, if the Boltzmann constant and μ are expressed in SI units $\bar{v}_{AB}$, $\bar{v}_{AA}$ will be in m s^{-1}, easily converted to cm s^{-1}.

(ii) Fraction of collisions, *F*, with sufficient energy for reaction

The SCT assumes molecules are simple hard spheres which can exchange kinetic energy along the line of centres on collision. The fraction of collisions, $F(\varepsilon)\mathrm{d}\varepsilon$, with relative kinetic energy lying between ε and $\varepsilon + \mathrm{d}\varepsilon$ is also given by the Maxwell–Boltzmann distribution law, from which:

$$F(\varepsilon)\mathrm{d}\varepsilon = \frac{1}{\bar{k}T}\mathrm{e}^{-\varepsilon/\bar{k}T}\,\mathrm{d}\varepsilon \tag{5.1.6}$$

If the minimum energy for reaction is ε_{min}, then by this model all collisions with energy $\varepsilon \geqslant \varepsilon_{\text{min}}$ result in reaction. The fraction of collisions with $\varepsilon \geqslant \varepsilon_{\text{min}}$ is

$$F = \int_{\varepsilon_{\text{min}}}^{\infty} F(\varepsilon)\,\mathrm{d}\varepsilon = \int_{\varepsilon_{\text{min}}}^{\infty} \frac{\mathrm{e}^{-\varepsilon/\bar{k}T}}{\bar{k}T}\,\mathrm{d}\varepsilon$$

and (5.1.7)

$$F = \mathrm{e}^{-\varepsilon_{\text{min}}/\bar{k}T}$$

In molar units, if L is the Avogadro constant then $L\varepsilon_{\text{min}} = E_{\text{min}}$ and $L\bar{k} = R$. Thus

$$F = \mathrm{e}^{-E_{\text{min}}/RT} \tag{5.1.8}$$

The rate constant

The rate of reaction on the SCT model as expressed in equation (5.1.1) is:

$$\text{Reaction rate} = Z_{AB} \times F$$

From (5.1.3) and (5.1.8)

$$\text{Reaction rate} = Z_{AB}{}^{0}\,\mathrm{e}^{-E_{\text{min}}/RT}\,C_A C_B \tag{5.1.9}$$

The experimental rate equation, with concentration units of molecule cm^{-3}, is

$$\text{Reaction rate} = k\,C_A C_B$$

Comparing terms with (5.1.9) gives the result for the SCT

$$k = Z_{AB}{}^{0}\,\mathrm{e}^{-E_{\text{min}}/RT}\ \text{molecule}^{-1}\,\text{cm}^{3}\,\text{s}^{-1} \tag{5.1.10}$$

Taking $Z_{AB}{}^{0} = \sigma_{AB}\bar{v}_{AB}$ and $\bar{v}_{AB}$ from (5.1.5) gives

$$k = \sigma_{AB}\left(\frac{8\bar{k}T}{\pi\mu}\right)^{1/2}\mathrm{e}^{-E_{\text{min}}/RT} \tag{5.1.11}$$

To convert k from units of molecule^{-1} to mol^{-1} the expression is multiplied by the Avogadro constant.

Comparison with experiment

The SCT result for the rate constant has the same form as the experimental Arrhenius equation—exponential dependence on energy and a pre-exponential factor almost independent of temperature. In fact $Z_{AB}{}^{0}$ contains a $T^{1/2}$ term, but this is too small a variation to be readily detected in experimental data. The relationship between the experimental activation energy and E_{min} can be shown explicitly by differentiating the logarithmic form of the Arrhenius equation.

$$RT^2\left(\frac{\mathrm{d}\,(\ln k)}{\mathrm{d}T}\right) = E_{\text{exp}} \tag{1.10.2}$$

The SCT result may be written

$$k = BT^{1/2}\, e^{-E_{min}/RT}$$

where $Z_{AB}{}^0 = BT^{1/2}$, and B is independent of temperature. So

$$RT^2\left(\frac{d\,(\ln k)}{dT}\right) = RT^2\left(\frac{1}{2T} + \frac{E_{min}}{RT^2}\right) \tag{5.1.12}$$

Comparing the right-hand sides of (1.10.2) and (5.1.12) gives

$$E_{exp} = E_{min} + \tfrac{1}{2}RT \tag{5.1.13}$$

A major defect in the theory is that the minimum energy for reaction cannot be calculated independently. To test the success of the theory in predicting pre-exponential factors E_{min} must be derived from the experimental activation energy as in (5.1.13). $Z_{AB}{}^0$ is a function of universal constants, the molecular weights of reactants and their cross sections which may be estimated from gas viscosity measurements, and so it is independently calculable. For typical elementary bimolecular reactions in the gas phase, the collision number $Z_{AB}{}^0$ lies in the range 3×10^{13}–10^{15} mol^{-1} cm^3 s^{-1}. Table 5.1 presents some experimental data. The measured A_{exp} values do not exceed $Z_{AB}{}^0$ and in some cases are in good agreement, particularly for reactions between atoms or simple radicals and molecules. However, for more complex species the pre-exponential factor is several orders of magnitude less than $Z_{AB}{}^0$. This is not at all surprising in view of the drastic assumption that reactant molecules are simple hard spheres. A popular rationalization of these discrepancies is to attribute them to a steric effect. For example, in the reaction:

$$A + BC \rightarrow AB + C$$

Table 5.1
Arrhenius parameters for bimolecular reactions arranged in order of decreasing pre-exponential factor.

Reaction	A_{exp}/mol^{-1} cm^3 s^{-1}	E_{exp}/kJ mol^{-1}
$O + N_2 \rightarrow NO + O$	10^{14}	315
$OH + H_2 \rightarrow H_2O + H$	8×10^{13}	42
$Cl + H_2 \rightarrow HCl + H$	8×10^{13}	23
$CH_3 + CH_3 \rightarrow C_2H_6$	2×10^{13}	~0
$F + CCl_4 \rightarrow FCl + CCl_3$	10^{13}	43
$NO + Cl_2 \rightarrow NOCl + Cl$	4×10^{12}	85
$CF_3 + CH_4 \rightarrow CF_3H + CH_3$	10^{12}	47
$SO + O_2 \rightarrow SO_2 + O$	3×10^{11}	27
$CH_3 + C_2H_6 \rightarrow CH_4 + C_2H_5$	2×10^{11}	44
$CH_3 + C_3H_6 \rightarrow C_4H_9$	10^{11}	~25
$C_6H_5 + H_2 \rightarrow C_6H_6 + H$	5×10^{10}	27
$O_3 + C_3H_8 \rightarrow C_3H_7O + HO_2$	10^9	51

if A and B are small and C is a very large group, then collisions in which A strikes B directly are far more likely to lead to reaction than if A approaches C which is efficiently shielding B. The steric effect is incorporated by writing

$$k = PZ^0\, e^{-E_{min}/RT}$$

where P is a steric factor whose value is given by A_{exp}/Z^0. P really represents an arbitrary correction factor and is not a useful contribution to quantitative rate theory.

5.2 A Modified Collision Theory

The SCT takes no account of the intramolecular detail of reactions, but when molecules gain energy on collision there must be some effect on internal energy modes. Electronic and rotational states are unlikely to be crucial since electronic excitation requires too much energy to play a major role and rotational quanta are too small to influence reactivity greatly. However, the distribution of energy among vibrational modes must be a critical factor since bond rupture and formation are of central importance in a reaction. It is worth exploring an extension to collision theory which takes account of the distribution of energy in vibrational modes, and the probability of a critical energy accumulating in the one mode that leads to reaction. The rate of reaction is still given by the product of collision frequency and fraction of collisions that result in reaction, but the latter term must be examined in more detail.

If energy can be distributed in n modes the fraction of species with total energy between ε and $\varepsilon + d\varepsilon$, irrespective of the way this energy is distributed between these modes, is

$$F(\varepsilon)\,d\varepsilon = \frac{1}{(n-1)!}\left(\frac{\varepsilon}{\bar{k}T}\right)^{n-1}\frac{e^{-\varepsilon/\bar{k}T}}{\bar{k}T}\,d\varepsilon \qquad (5.2.1)$$

which should be compared with the SCT expression, (5.1.6).

Whether molecules with total energy ε react depends on the probability of a minimum amount of energy, ε_{min}, being in the critical mode. Of course if $\varepsilon < \varepsilon_{min}$ then reaction is impossible. If $\varepsilon \geqslant \varepsilon_{min}$ the probability $P(\varepsilon)$ of having at least ε_{min} in the critical mode is:

$$P(\varepsilon) = \left(\frac{\varepsilon - \varepsilon_{min}}{\varepsilon}\right)^{n-1} \qquad (5.2.2)$$

Thus the higher the energy, ε, the greater the probability of reaction. This variation of probability with energy is illustrated in figure 5.2 for different values of n. The diagram also shows the SCT hard-sphere assumption which corresponds to $P = 0$ for $\varepsilon < \varepsilon_{min}$ and $P = 1$ for $\varepsilon \geqslant \varepsilon_{min}$.

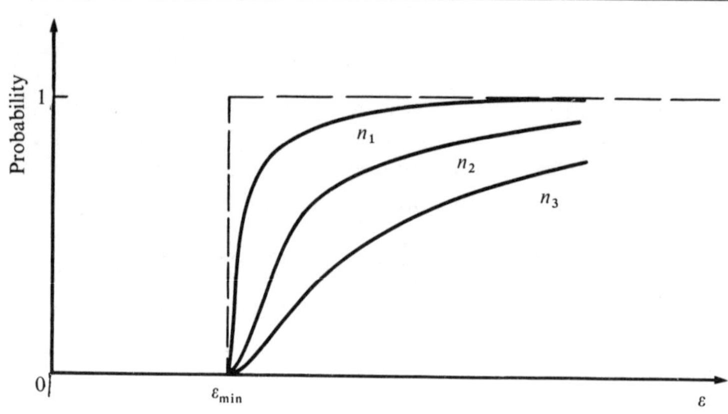

Figure 5.2
Variation of probability of reaction with energy. Broken line, SCT; $n_3 > n_2 > n_1$.

F, the fraction of collisions resulting in reaction, combines the fraction of molecules having a given energy and the probability of reaction at that energy:

$$F = \int_{\varepsilon_{\min}}^{\infty} P(\varepsilon)F(\varepsilon)\,\mathrm{d}\varepsilon \tag{5.2.3}$$

Reaction rate = collision frequency × F and so:

$$\text{Reaction rate} = Z_{AB}{}^{0}C_{A}C_{B}\int_{\varepsilon_{\min}}^{\infty} F(\varepsilon)P(\varepsilon)\,\mathrm{d}\varepsilon$$

From (5.2.1) and (5.2.2)

$$\text{Reaction rate} = Z_{AB}{}^{0}C_{A}C_{B}\int_{\varepsilon_{\min}}^{\infty} \frac{1}{(n-1)!}\left(\frac{\varepsilon}{\bar{k}T}\right)^{n-1}\frac{\mathrm{e}^{-\varepsilon/\bar{k}T}}{\bar{k}T}\left(\frac{\varepsilon-\varepsilon_{\min}}{\varepsilon}\right)^{n-1}\mathrm{d}\varepsilon$$

which integrates to

$$\text{Reaction rate} = Z_{AB}{}^{0}C_{A}C_{B}\,\mathrm{e}^{-\varepsilon_{\min}/\bar{k}T}$$

So, paradoxically, this gives the same result as the SCT. The model which allows distribution of energy among n modes but limits reaction to the case where the entire $\varepsilon_{\min}$ must concentrate in one critical mode, turns out as restrictive as the hard-sphere model in this application. It does, however, give a preliminary insight that will be useful in other contexts, such as the detailed treatment of unimolecular reactions (sections 5.10, 11).

5.3 Potential Energy Surfaces

The SCT was established before successful quantum mechanical interpretations of molecular structure stimulated a more detailed look at energy changes throughout a reactive collision. The potential energy of the diatomic system: $A + B \rightarrow AB$ was discussed in section 2. This case is straightforward, energy varies with $A \ldots B$ distance and so can be portrayed on a two-dimensional diagram. With a triatomic system, such as: $A + BC \rightarrow AB + C$, the simplest of wide kinetic interest, the situation becomes much more complicated. Energy varies with three parameters, say $A \ldots B$ and $B \ldots C$ distances and the angle $\widehat{ABC}$, and the complete energy diagram is four dimensional. If one parameter is fixed the energy calculations are simplified and the energy diagram is three dimensional and quite easily visualized. This can be achieved by the reasonable postulate that reaction is most probable for a *line-of-centres collision*, i.e. at a fixed $\widehat{ABC}$ angle of 180°. When A approaches B along the $B \ldots C$ axis, the situation is sterically efficient, and interaction of A with C, which would raise the total energy, is minimized. Plotting the energy of the triatomic system for various $A \ldots B$ and $B \ldots C$ distances gives a *potential energy surface* or *hypersurface* as shown in figure 5.3, which attempts a perspective view.

The unlined face of the diagram is a section through the surface perpendicular to the $B \ldots C$ axis. This face gives the potential energy curve for molecule AB at a large $B \ldots C$ separation. The equilibrium internuclear distance for the molecule is

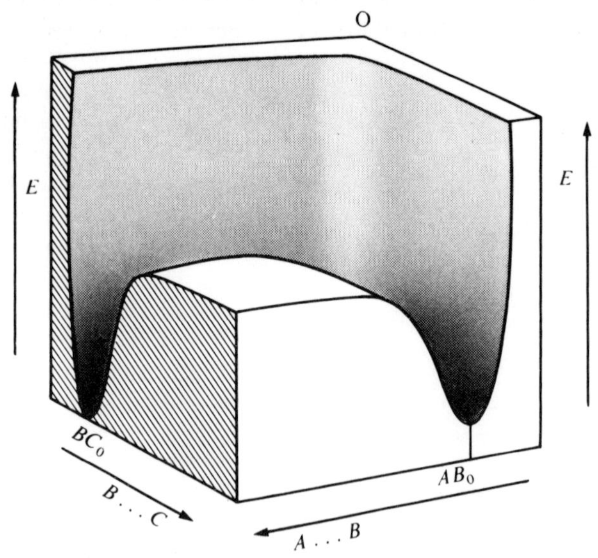

Figure 5.3
Potential energy surface for the system $A + BC \rightarrow AB + C$. The origin from which internuclear distance is measured is at O.

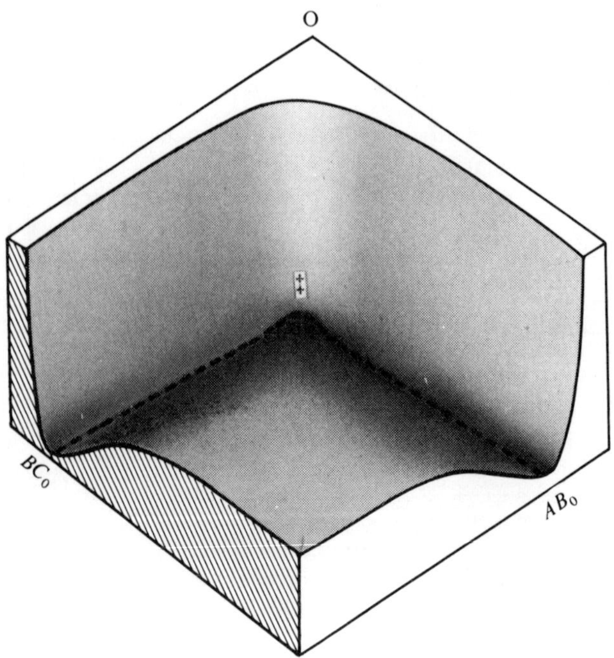

Figure 5.4
Potential energy surface for the system $A + BC \rightarrow AB + C$ showing the col, ‡, and the reaction path ---→.

indicated as AB_0. Similarly the lined face gives the potential energy curve for BC at large $A \ldots B$ separation. During the course of the reaction, $A + BC \rightarrow AB + C$, the system must move through point BC_0 to point AB_0. The most probable route has the lowest overall potential energy, and this is shown more clearly if the diagram is tilted toward the reader as in figure 5.4. The reaction follows the path indicated by the dashed line, along what may be called a reactant valley and, pursuing the geographical analogy, over a saddle point or col, marked ‡, into the product valley. The plan of the potential energy surface can be plotted as a contour diagram where the contour lines join points of equal potential energy. Figure 5.5 clearly reveals the 'geographical' features: reactant and product valleys separated by the col; and also a plateau of constant energy where $A \ldots B$ and $B \ldots C$ are large—where the system consists of

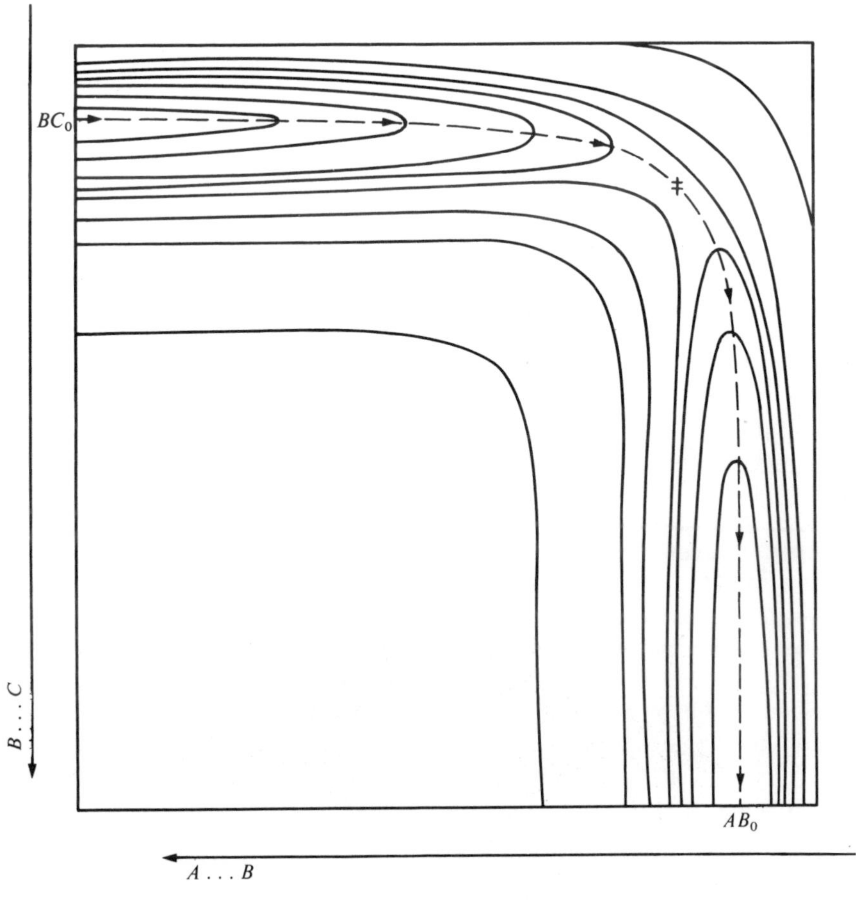

Figure 5.5
Contour diagram of potential energy surface for system $A + BC \rightarrow AB + C$ showing the col, ‡, and reaction path ---→. In this example the contours show that the product valley is at a higher energy than the reactant valley—the reaction is endothermic.

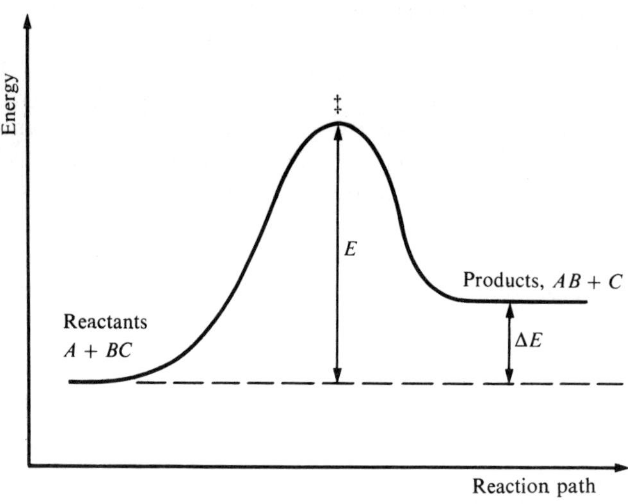

Figure 5.6
Variation of energy along reaction path for $A + BC \rightarrow AB + C$.

separate atoms. The most probable reaction path is again indicated, and it is instructive to plot how energy varies along this path as in figure 5.6.

In this particular example the product valley is at higher energy than the reactant valley—the reaction is endothermic by ΔE. The energy barrier, height E, stands out and clearly the point at the top of the barrier is crucially important. Although it is the maximum energy along the reaction path, figure 5.5 shows that it has the minimum energy for motion through the col at right angles to the path. There is a definite, if entirely transitory, molecular configuration at the col $(A \ldots B \ldots C)^{\ddagger}$ often called the activated complex. These features form the basis for activated-complex theory of the rate constant discussed in section 5.5. It must be emphasized that the height of the energy barrier is not identical with the activation energy measured experimentally, though their values are usually similar in magnitude. The surface described is for the lowest electronic energy state of the system neglecting zero-point energy, which would be represented by another surface above that shown. Potential energy surfaces for reactions involving more than three atoms require the corresponding extra dimensions. They are hypersurfaces in multidimensional space and are extremely difficult to treat. It is sometimes possible to approximate groups of atoms to mass points, and so achieve a useful, effectively triatomic system.

5.4 Potential Energy Calculations

This section requires a greater knowledge of molecular theory than the rest of the chapter and it may be omitted at this stage. The material is also relevant to the discussion in chapters 8, 9, 10.

Most attention has been devoted to the prototype three-electron system: $H + H_2 \rightarrow H_2 + H$; or, if we label each atom:

$$H_A + H_BH_C \rightarrow H_AH_B + H_C \qquad (5.4.1)$$

which can be studied experimentally by the rate of para-ortho hydrogen conversion. Perhaps the best fit to the Arrhenius equation is given by the experimental activation energy 31 ± 4 kJ mol^{-1}.[3] Calculations confirm that the line-of-centres collision is the lowest energy path and, with this restriction, surfaces of the type discussed above can be calculated. Even for this simple prototype reaction it is difficult to achieve a surface that is useful to the kineticist. His main concern is with the magnitude of the energy barrier to reaction, which should not differ greatly from the experimental activation energy. The energy barrier is the difference between the energy of the reactant valley and of the activated-complex configuration $(H \ldots H \ldots H)^{\ddagger}$ at the col. Quantum mechanical calculations give these energies relative to the common reference state of three ionized atoms. The resulting energies are very large, about 50 eV (4750 kJ mol^{-1}), but their difference, the energy barrier, is only about 0·4 eV. Hence to calculate the barrier height with 10% accuracy requires a precision of better than 0·1% in the energy calculations, a tall order even for the three-electron case and as yet unobtainable for more complicated systems. Theoretical kineticists have attempted approximate purely quantum mechanical methods but much effort has been put into semi-empirical methods and even empirical approaches can be useful.

Purely quantum mechanical methods

The pioneering effort was based on an approximate equation proposed by *London*[4] who described the energy of the triatomic system by a combination of terms for the various possible diatomics H_AH_B, H_BH_C, H_AH_C. The energy terms were those used in the Heitler–London method[5] for H_2 which gives the energy of the diatomic molecule as $E = Q \pm J/(1 + S^2)$ where Q, J, and S are integrals of combined 1s atomic wavefunctions (orbitals). Q is the *Coulombic*, J the *exchange*, and S the *overlap* term. In the more approximate equation neglecting overlap, $E = Q \pm J$. Here $Q + J$ is the energy of the H_2 bonding molecular orbital, $Q - J$ that of the H_2 antibonding orbital.

In the triatomic system the energy is given by a combination of such diatomic terms. If the Coulombic and exchange terms for H_BH_C are A and α, and the corresponding terms for H_AH_C are B and β, and for H_AH_B they are C and γ, London's equation for the energy, E, of the triatomic system is:

$$E = A + B + C \pm [\tfrac{1}{2}\{(\alpha - \beta)^2 + (\beta - \gamma)^2 + (\gamma - \alpha)^2\}]^{1/2} \qquad (5.4.2)$$

London's method is historically important but the energies obtained are not reliable and the surface shows a remarkable additional feature—a 'basin' or 'well' at the top of the energy barrier. This has a physical significance, it denotes a slightly stabilized molecular entity—a real collision complex—rather than an activated complex, which is a hypothetical species at a point (the col) on the continuous transition from reactants to products. Experimental evidence and more refined calculations show that this feature of primitive calculations is in error. Many modifications of London's method have been suggested, but more success has attended the use of *variation methods*.

[3] Discussed in K. J. Laidler, *Theories of Chemical Reaction Rates*, McGraw-Hill, 1969.
[4] F. London, *Z. Elektrochem.*, **35**, 552 (1929).
[5] H. Heitler and F. London, *Z. Phys.*, **44**, 455 (1927).

This important principle in quantum mechanical calculations can be stated simply. If the wave equation for a system is $H\psi = E\psi$ where H is the Hamiltonian operator and E the energy, then if one selects an approximate wavefunction ϕ, the energy given by $\oint \phi H \phi^* \, d\tau$ must be above the true energy. Thus if the chosen wavefunction is varied systematically, the lowest energy that results denotes the best approximation. For example, H.H.H may be represented by a suitable combination of 1s orbitals, and the effective nuclear charge and the polarity of the combined orbitals varied to give the best energy at each set of internuclear distances. Such attempts usually result in barriers that are too high, about 80 kJ mol^{-1} at best, but with improving computer technology it has become possible to start with more complex theoretical functions. Including p as well as s orbitals in the calculations has given barrier heights closely approaching the barrier height expected from experiment. Despite such improvements for H.H.H, accurate calculations for more complex reactants are still difficult. However, one of the most important trends in applied quantum chemistry is the development of more feasible schemes for few-electron triatomic systems such as F.H.H. Extending these methods beyond few-electron systems still presents great conceptual problems.

Semi-empirical methods

These adjust the theory to fit known patterns of experimental results and hopefully lead to a consistent approach for a wider range of reactions. The earliest attempt was by Eyring and Polanyi based on the London equation, and popularly designated the LEP method[6]. The triatomic system was expressed as a combination of diatomic terms as in equation (5.4.2), and for each diatomic term the variation of energy with internuclear distance, r, was given by the empirical Morse equation:

$$E = D(1 - e^{-x(r-r_e)})^2 \qquad (5.4.3)$$

D is the classical dissociation energy, r_e the equilibrium internuclear distance and x is a constant. For the relative contribution of Coulombic and exchange terms they chose 15% Coulombic. Later theoreticians have tried a series of Coulombic contributions, calculating the corresponding barrier heights. Results vary from 80 kJ mol^{-1} at 7% to 30 kJ mol^{-1} at 20%. The LEP method also gave a basin at the top of the barrier that only disappears with a Coulombic contribution greater than 30%.

The best subsequent attempt, giving perhaps the most reliable surface for this reaction, was that of Porter and Karplus[7]. They expressed the Coulombic contribution again as the sum of three diatomic terms, but three-atom exchange terms were used, and double exchange integrals and overlap terms were added. The resulting surface exhibited no basin and the barrier height was 36 kJ mol^{-1}. Part of Karplus and Porter's surface is shown in figure 5.7.

In recent years surfaces for reactions other than $H + H_2$ have received increased attention as molecular dynamics theory has grown (see chapter 8). Experimental results from molecular beam and chemiluminescence studies (as discussed in chapters 9, 10) have provided complementary insights into surfaces for a widening range of reactions. *Exothermic reactions* of the type $A + BC \rightarrow AB + C$, where B may be a

[6] H. Eyring and M. Polanyi, *Z. Phys. Chem., Lpz.*, **B12**, 279 (1931).
[7] R. N. Porter and M. Karplus, *J. Chem. Phys.*, **44**, 1105 (1964).

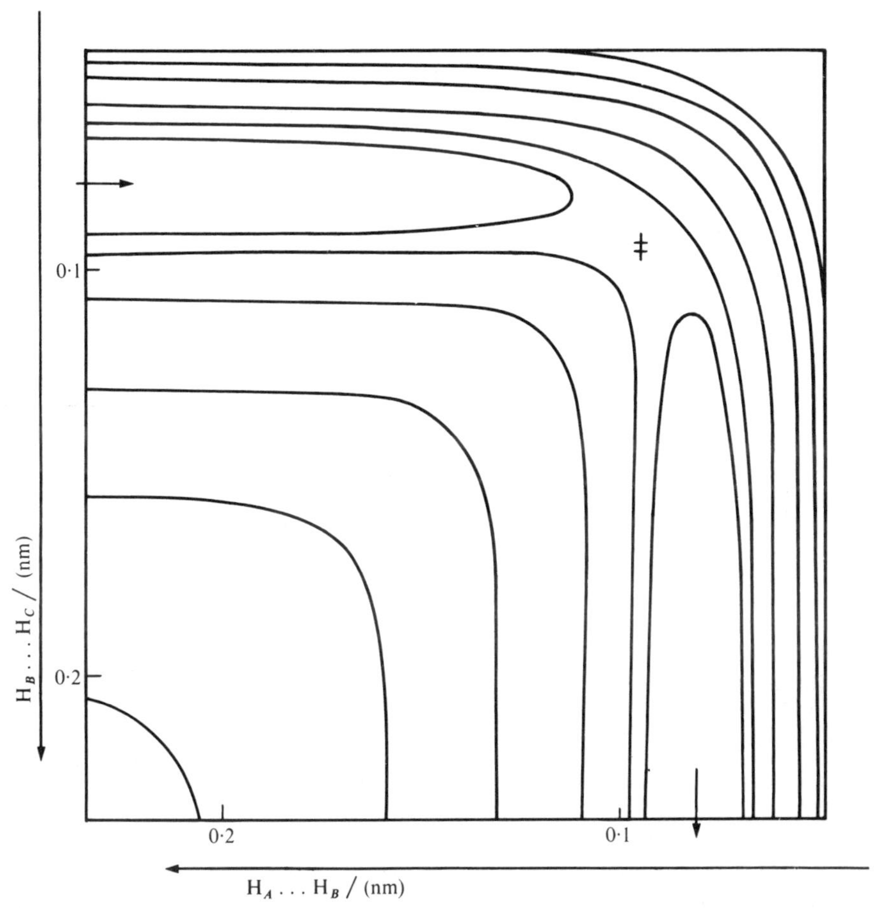

Figure 5.7
Semi-empirical potential energy surface for $H_A + H_BH_C \rightarrow H_AH_B + H_C$.

group of atoms, have been investigated in some detail. They can differ widely in the fraction of energy released that appears as internal energy of products (especially vibrational energy) rather than translational energy. This behaviour can be correlated with different types of surface.

Attractive surfaces

An illustrative example is given in figure 5.8. The activated complex at the col forms at rather long *A . . . B* distance, whilst the *B*—*C* bond is relatively undisturbed. The exothermicity of the reaction is released as the system moves 'downhill' from the col into the product valley. Thus, in this case, the energy is released as the new bond is formed, and it is retained as vibrational excitation in *A*—*B* when atom *C* separates.

Repulsive surfaces

As shown in figure 5.9, the activated complex at the col is situated where the *A . . . B* distance is close to the *A*—*B* bond length in the product, and *B . . . C* is increasing

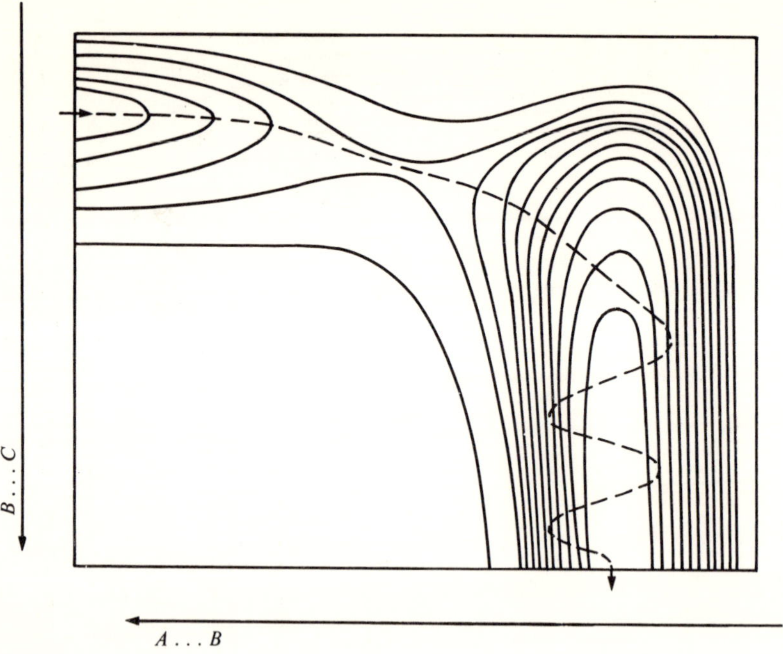

Figure 5.8
An exothermic reaction $A + BC \rightarrow AB + C$ with an attractive potential energy surface. Vibrational excitation is represented schematically on the reaction path.

as atom C separates. The downhill slope occurs along the coordinate for $B \ldots C$ separation, and energy is released with the A—B product virtually formed and atom C moving away. Hence a high fraction of exothermicity appears as translational energy of the products.

Mixed energy release

Some energy release usually occurs in the curved section of the reaction path. Here both $A \ldots B$ and $B \ldots C$ distances are changing and there is mixed energy release into both vibration and translation. Thus some vibrational excitation occurs even on repulsive surfaces. The efficiency of converting exothermicity into vibrational excitation increases as the surface becomes increasingly attractive.

The lowest degree of vibrational excitation is found with repulsive surfaces when reactant A is a very light atom—the *light atom anomaly*[8].

Empirical methods

There have been efforts to produce empirical surfaces, for example the Morse curve (see equation (5.4.3)) for a diatomic can be rotated through 90° first about a point

[8] There is some dispute over the existence of this effect, see for example, D. L. Bunker and C. A. Parr, *J. Chem. Phys.*, **52**, 5700 (1970).

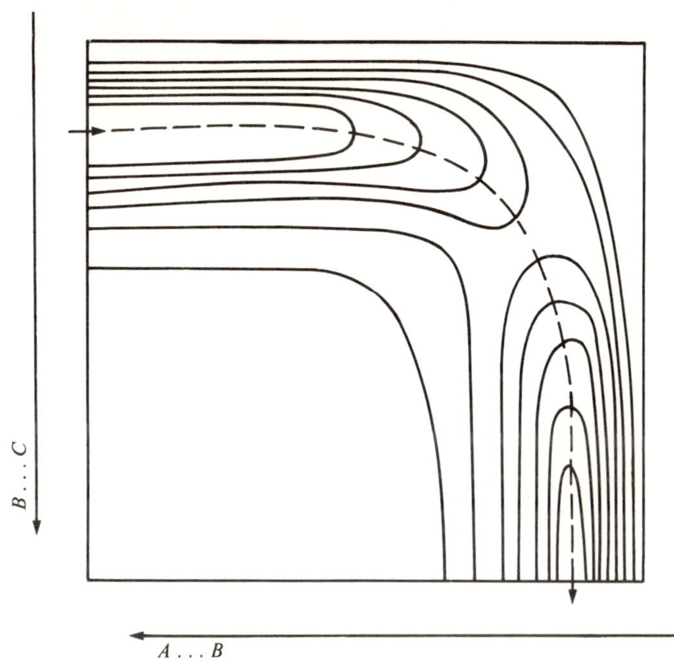

Figure 5.9
An exothermic reaction $A + BC \rightarrow AB + C$ with a repulsive potential energy surface.

corresponding to large $A \ldots B$ distance, then about a point at large $B \ldots C$ distance. On rotation the curve is systematically distorted giving the rise to the col between the two valleys[9].

Bond energy–bond order (BEBO)

The BEBO method[10] calculates the potential energy changes along the reaction path only. In the $A \ldots B \ldots C$ complex there are two 'partial' bonds, or bonds of fractional order—in line with Pauling's concept. The method assumes that the most probable reaction path is where the sum of bond orders is unity throughout. For each partial bond the interatomic distance, r, when its bond order is n, is given by the empirical relationship: $r = r_0 - \text{constant} \times \ln n$, where r_0 is the single-bond length for the particular pair of atoms. The energy of the partial bond is $E = Dn^p$ where D is the single bond energy and p a factor characteristic of the bond, evaluated empirically from relations between bond energy, length and order. The BEBO method thus uses empirical relations from outside the field of kinetics. It immediately gives the value of n for the barrier, the barrier height and the atomic configuration of the activated complex. When applied to hydrogen transfer reactions ($B \equiv \text{H}$) it has been very successful in predicting barrier heights close to the experimental activation energy.

[9] e.g. F. T. Wall and R. N. Porter, *J. Chem. Phys.*, **36**, 3256 (1962).
[10] H. S. Johnston, *Gas Phase Reaction Rate Theory*, p. 177, Ronald Press Company, 1966.

Empirical activation energies

Some useful correlations have been obtained between activation energy and heat of reaction for exothermic reactions. The first popular example was the linear relationship proposed by Evans and Polanyi[11]:

$$E_{exp} = \alpha \Delta H + c$$

where c and α are empirical constants for a related series of reactions, the constants differing for different series. ΔH is the heat of reaction, negative for these exothermic cases. This approach was extended to a wide range of abstraction reactions by Semenov[12]. For homologous series of reactions such relationships hold very well as shown in figure 5.10.

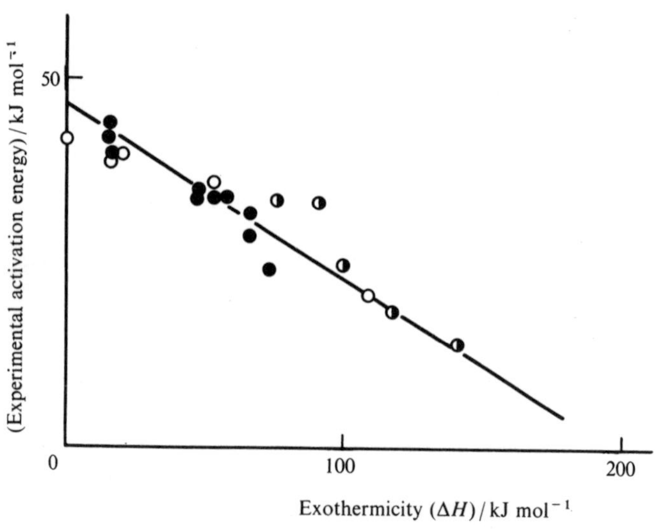

Figure 5.10
Plot of an empirical relationship for activation energies: $E_{exp} = -0{\cdot}25\ \Delta H + 48$ for the reaction series: ○, $H + RH \rightarrow H_2 + R$; ●, $CH_3 + RH \rightarrow CH_4 + R$; ◑, $H + RCl \rightarrow HCl + R$.

5.5 Activated-Complex Theory for Bimolecular Reactions

The construction of potential energy surfaces stimulated interest in a more detailed picture of reactive collisions than the SCT could provide. Attention was focused on the transient molecular configuration at the top of the energy barrier and in 1935 a new method of calculating rate constants was suggested independently by Evans and Polanyi and by Eyring[13]. Eyring originally called the molecular species at the col the *activated complex* whereas Polanyi used the term *transition state*. Many authors have

[11] M. G. Evans and M. Polanyi, *Trans. Faraday Soc.*, **34**, 11 (1938).
[12] N. N. Semenov, *Some Problems in Chemical Kinetics and Reactivity*, Pergamon, 1958.
[13] H. Eyring, *J. Chem. Phys.*, **3**, 107 (1935); M. G. Evans and M. Polanyi, *Trans. Faraday Soc.*, **31**, 875 (1935).

since made a formal distinction between the two terms, the system at the col being in the transition state whereas the corresponding complex of atoms is the activated complex. The distinction does not seem helpful and the latter term is retained here. Similarly the theory itself has been called Transition-State Theory, Activated-Complex Theory and even Absolute-Rate Theory. The last designation appears rather presumptuous and of the other two, Activated-Complex Theory (ACT) is preferred here.

The theory postulates a state of equilibrium between reactants and the activated complex. Though the complex has only transient existence—inevitably breaking down to give products—it is regarded as a thermodynamic entity and the equilibrium is treated by the standard methods of statistical mechanics*. The reaction rate is then given by the concentration of activated complexes multiplied by their decomposition frequency. By this relatively simple device an immense number of rate constants have been calculated with considerable success for the resulting pre-exponential factors. There is a clear relationship between the experimental activation energy and the energy barrier on the surface, but it must be remembered that for most reactions the energy barrier has not been calculated.

The reaction path

This is the route of minimum overall energy from reactants to products over the potential energy surface for the reaction. The variation of energy along the reaction path in a typical case is shown in figure 5.11. The energy surface is the 'classical' surface since it does not include quantization of the species. For reactants in their ground electronic states, rotational energy makes a comparatively small contribution to reactivity and may be neglected, but vibrational energy must be taken into account. Except at very high temperatures the species may be assumed to occupy their ground vibrational states which means adding zero-point energies to the classical energies. Thus, as shown in figure 5.11, the effective energy barrier to reaction is the difference in energy, $E_0^{\ddagger}$, between these ground states of reactant and activated complex.

The equilibrium hypothesis

Species, originally reactants, which reach the top of the energy barrier become activated complexes and pass on to become products. There is no reflection back to reactants from the col, even though equilibrium statistical mechanics is applied to the relationship between reactants and the activated complexes. This is the heart of the conceptual difficulty experienced by many students. Although the usefulness of the hypothesis can be justified by the widespread success of calculations based upon it, a more detailed examination does provide further support.

Consider a system with reactants and products in equilibrium. The rates of forward and back reaction are equal, and both proceed over the col and so through the activated complex. Distinguishing formally between activated complexes formed in the forward reaction, $\ddagger_f$, and the back reaction, $\ddagger_b$ we have:

Reactants $\rightarrow \ddagger_f \rightarrow$ products

Reactants $\leftarrow \ddagger_b \leftarrow$ products

* The Statistical Mechanics employed in this chapter is of the elementary level given in standard general texts on Physical Chemistry such as: W. J. Moore, *Physical Chemistry*, Longman, 1972; or G. M. Barrow, *Physical Chemistry*, McGraw-Hill, 1973.

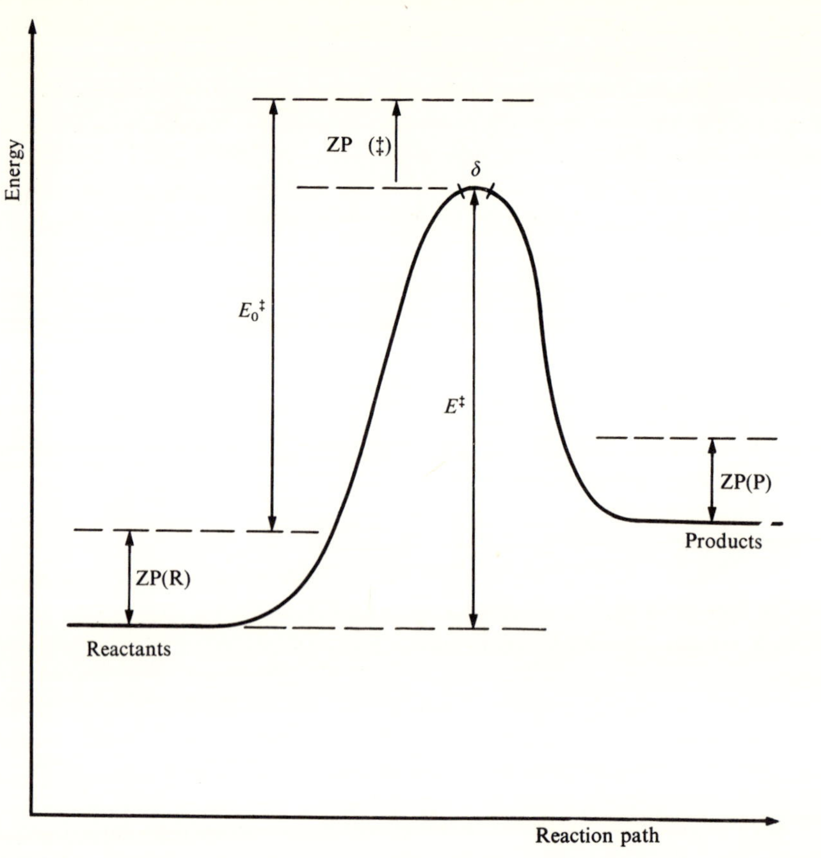

Figure 5.11
Variation of energy along the reaction path on the 'classical' surface (continuous line). $E^‡$ is the 'classical' barrier height. ZP (R, P, ‡) represent the zero-point energies of reactants, products, and activated complex respectively. $E_0{}^‡$ is the difference in energy, reactants → complex required by the ACT. δ is the distance along reaction path for which activated complex exists (see derivation 2).

In this situation, where complete equilibrium exists, it is certainly justifiable to calculate the concentration of activated complexes by equilibrium statistical mechanics. With the rates of forward and back reactions equal, the two types of complex will be in equal concentration. As a 'thought experiment', imagine all products suddenly removed. For the given reactant concentration, the rate of the forward reaction must be the same whatever the rate of the back reaction. Thus the relationship between reactant and activated-complex concentration deduced for full equilibrium is still valid, and statistical mechanics can be applied to reaction in one direction[14].

Some very detailed calculations have been carried out using non-equilibrium statistical mechanics and these indicate that the equilibrium approach gives satisfactory results except where the activation energy is very low. The equilibrium hypothesis implies a

[14] For a very interesting critique of the model see D. L. Bunker, *Acct. Chem. Res.*, **7**, 195 (1974).

Boltzmann energy distribution for species present and this distribution may be seriously distorted where the activation energy is very low, since the consequent, very high reaction rate may result in serious depletion of higher energy reactants. A limit of $E/RT < 5$ has been suggested for the satisfactory application of the method, although the results for the ACT seem quite successful even for these low activation energy reactants.

Derivation of the rate equation for bimolecular reactions

For the reaction:

$$A + B \rightarrow \text{products}$$

the basic model of the ACT is:

$$A + B \rightleftharpoons \ddagger \rightarrow \text{products} \qquad (5.5.1)$$

that is, an equilibrium is postulated between reactants and an activated complex which decomposes to products. The reaction rate is given:

$$\text{Reaction rate} = \begin{pmatrix}\text{concentration}\\ \text{of } \ddagger\end{pmatrix} \times \begin{pmatrix}\text{decomposition}\\ \text{frequency for } \ddagger\end{pmatrix}$$

$$= [\ddagger] \times f \qquad (5.5.2)$$

The concentration of the activated complex can be expressed in terms of the equilibrium constant:

$$K^{\ddagger} = [\ddagger]/[A][B]$$

whence

$$[\ddagger] = K^{\ddagger}[A][B]$$

and

$$\text{Reaction rate} = fK^{\ddagger}[A][B] \qquad (5.5.3)$$

From this point the derivation can proceed in various ways, two examples of which will be given here.

Derivation 1

Consider the motion over the top of the energy barrier as a very weak vibration with frequency $\nu^{\ddagger}$. This corresponds to the existence of a very shallow depression at the col. The decomposition frequency is then equal to this vibrational frequency in the critical mode that leads to decomposition. Thus:

$$f = \nu^{\ddagger}$$

and from equation (5.5.3)

$$\text{Reaction rate} = \nu^{\ddagger}K^{\ddagger}[A][B] \qquad (5.5.4)$$

The standard methods of statistical mechanics give the following expression for the equilibrium constant:

$$K^{\ddagger} = \frac{Q_{\ddagger}}{Q_A Q_B}\, e^{-E_0^{\ddagger}/RT} \qquad (5.5.5)$$

$E_0^\ddagger$ is the energy barrier to reaction illustrated in figure 5.11. $Q_\ddagger$, Q_A, and Q_B are the *molecular partition functions per unit volume* for the complex, A and B respectively. The molecular partition function of a species is the product of separate partition functions for *each* degree of translational, rotational and vibrational freedom, q_t, q_r and q_v respectively. Examples of partition functions are given in table 5.2, and:

$$Q = \prod_t q_t \prod_r q_r \prod_v q_v$$

Table 5.2
Some partition functions: m = molecular mass; I = moment of inertia for linear molecule; I_A, I_B, I_C = moments of inertia for non-linear molecule; ν = vibrational frequency (normal mode).

Type of energy	Partition function[a]	Order of magnitude
Translational, 1 degree of freedom	$q_t = (2\pi m\bar{k}T)^{1/2}/h$ per unit length	$\sim 10^8\ \text{cm}^{-1}$
Complete translational, 3 degrees of freedom	$q_t^3 = (2\pi m\bar{k}T)^{3/2}/h^3$ per unit volume	$\sim 10^{24}\ \text{cm}^{-3}$
Complete rotational[b] for linear molecule, 2 degrees of freedom	$q_r^2 = 8\pi^2 I\bar{k}T/h^2$	$\sim 10^2$
Complete rotational[b] for non-linear molecule, 3 degrees of freedom	$q_r^3 = \{\pi^{1/2}(8\pi^2\bar{k}t)^{3/2}/h^3\}(I_A I_B I_C)^{1/2}$	$\sim 10^3$
Vibrational, 1 degree of freedom (normal mode)	$q_v = (1 - e^{-h\nu/\bar{k}T})^{-1}$	$\sim$1–10

[a] Partition function for each degree of freedom is defined $q = \sum_{i=0}^{\infty} \omega_i\, e^{-\varepsilon_i/\bar{k}T}$ where there are ω_i energy levels of energy ε_i in the degree of freedom.

Total molecular partition function is the product of partition function for all degrees of freedom $Q = \Pi_i q_i$.

[b] For calculating reaction equilibrium constants, symmetry numbers must be included, but for ACT other procedures are preferable (see text, p. 93).

For completeness electronic partition functions should be included. However, where only ground electronic states are involved the electronic partition functions are unity. The most important exceptions are where the ground state is degenerate (e.g. $O_2(^3\Sigma_g^-)NO(^2\Pi)$) and electronic partition functions must be included in these cases.

There is a particular anomaly with the activated complex. Its molecular partition function is the product of individual partition functions for the various degrees of freedom as in a normal molecule, but with one exception. The vibrational mode corresponding to decomposition is considered an extremely weak vibration, leading to a useful approximation for the partition function. Let the molecular partition function be:

$$Q_\ddagger = Q_\ddagger' q_v^\ddagger$$

$Q_\ddagger'$ is the molecular partition function from which one vibrational term is missing, and $q_v^\ddagger$ is that vibrational term, for the critical mode. From table 5.2

$$q_v^\ddagger = (1 - e^{-h\nu\ddagger/\bar{k}T})^{-1}$$

but for a very weak vibration, the quantum $h\nu^\ddagger$ is small and $\bar{k}T \gg h\nu^\ddagger$. This gives, to a good approximation:

$$q_v^\ddagger = \frac{\bar{k}T}{h\nu^\ddagger} \quad \text{and} \quad Q_\ddagger = Q_\ddagger' \frac{\bar{k}T}{h\nu^\ddagger} \tag{5.5.6}$$

substituting (5.5.6) into (5.5.5) gives

$$K^\ddagger = \frac{\bar{k}T}{h\nu^\ddagger} \frac{Q_\ddagger'}{Q_A Q_B} e^{-E_0^\ddagger/RT} \tag{5.5.7}$$

and substituting (5.5.7) into the rate expression (5.5.4) gives:

$$\text{Reaction rate} = \frac{\bar{k}T}{h} \frac{Q_\ddagger'}{Q_A Q_B} e^{-E_0^\ddagger/RT} [A][B] \tag{5.5.8}$$

The rather uncertain term $\nu^\ddagger$ thus cancels out. Comparison of (5.5.8) with the experimental rate equation:

$$\text{Reaction rate} = k[A][B]$$

gives the result:

$$k = \frac{\bar{k}T}{h} \frac{Q_\ddagger'}{Q_A Q_B} e^{-E_0^\ddagger/RT} \tag{5.5.9}$$

Derivation 2

Eyring's approach was to consider a smooth transition across the col, and that the activated complex exists over a distance δ along the reaction path at the col, as shown in figure 5.11. In this segment of the reaction path, what should be a vibrational mode of the complex is transformed into translational motion, and the decomposition rate for the complex is the speed at which distance δ is traversed. Again the molecular partition function for the complex can be expressed with the partition function for the anomalous mode separated:

$$Q_\ddagger = Q_\ddagger' q_v^\ddagger$$

In this derivation the vibrational partition function $q_v^\ddagger$ is replaced by a translational function $q_t^\ddagger$ and:

$$Q_\ddagger = Q_\ddagger' q_t^\ddagger$$

The translational partition function for motion in *one* dimension in a system, length δ, is:

$$q_t^\ddagger = (2\pi m^\ddagger \bar{k}T)^{1/2} \delta / h$$

and substituting into (5.5.5) gives:

$$K^\ddagger = (2\pi m^\ddagger \bar{k}T)^{1/2} \frac{\delta}{h} \frac{Q_\ddagger'}{Q_A Q_B} e^{-E_0^\ddagger/RT} \tag{5.5.10}$$

The frequency of the complex's decomposition, f, is given by $f = \bar{v}_\ddagger / \delta$ where $\bar{v}_\ddagger$ is the average speed with which complexes move through segment δ of the reaction path. From the Kinetic Molecular Theory for motion in one dimension

$$\bar{v}_\ddagger = \left(\frac{\bar{k}T}{2\pi m^\ddagger}\right)^{1/2} \quad \text{and} \quad f = \frac{\bar{v}_\ddagger}{\delta} = \frac{1}{\delta}\left(\frac{\bar{k}T}{2\pi m^\ddagger}\right)^{1/2} \tag{5.5.11}$$

Substituting (5.5.10) and (5.5.11) into the rate expression (5.5.3) gives

$$\text{Reaction rate} = fK^{\ddagger}[A][B]$$

$$= \frac{1}{\delta}\left(\frac{\bar{k}T}{2\pi m^{\ddagger}}\right)^{1/2} (2\pi m^{\ddagger}\bar{k}T)^{1/2} \frac{\delta}{h} \frac{Q_{\ddagger}'}{Q_A Q_B} e^{-E_0^{\ddagger}/RT}[A][B]$$

giving

$$\text{Reaction rate} = \frac{\bar{k}T}{h} \frac{Q_{\ddagger}'}{Q_A Q_B} e^{-E_0^{\ddagger}/RT}[A][B]$$

The uncertain term δ cancels out and the result is the same as equation (5.5.8) resulting from derivation 1. Thus both approaches give the same expression, (5.5.9), for the bimolecular rate constant. With the basic form of the theoretical expression established, a few extra factors must be mentioned as they result in the modification of the expression in some cases.

The transmission coefficient

This is a correction factor usually symbolized κ to allow for the possibility that not every activated complex is transformed into the particular product specified in the reaction. Some other product may be formed, or the complex reflected back to reactants for some reason. Occasionally κ seems to be used as an empirical adjustment to achieve agreement with experiment, but this is not a satisfactory procedure. The transmission coefficient is incorporated in the theory by multiplying the result for the rate constant (the right-hand side of equation (5.5.9)) by κ, whose value must lie between 0 and 1. Three specific cases for $\kappa < 1$ will be outlined:

(i) Some reactions are known to proceed via an identifiable collision complex of finite lifetime associated with a basin in the potential energy surface. The complex has a lifetime corresponding to several molecular rotations, and there is a much enhanced possibility of returning to reactants. Such collision complexes are *not* activated complexes in the terms of the theory, but the reduced reaction rate may be expressed as a transmission coefficient of less than unity.

(ii) The combination of atoms:

$$A + A \rightarrow A_2$$

gives a diatomic molecule, where the bond initially contains all the energy released when it is formed. This is, of course, the bond dissociation energy, and the new molecule will dissociate in its first vibration unless some energy is lost to a third body on collision. Thus the simple bimolecular reaction as written will have a transmission coefficient of zero, and atomic recombination reactions are always termolecular.

(iii) In some reactions the potential energy surface is crossed in the region of the col by another surface. There is the formal possibility of the system crossing to the second surface, and not giving the reaction in question. If the probability of crossing is low, the reaction still proceeds mainly to the required products and the transmission coefficient is near unity. If the crossing probability is high, the transmission coefficient for the original reaction is very low.

A special case of a reaction with a low transmission coefficient is where the requisite products have a different total spin multiplicity from the reactants. That systems resist any change in spin multiplicity is known as the *Wigner spin conservation rule*. An example of such a low probability process is, for species in their electronic ground states:

$$O(^3P) + CO(^1\Sigma) \rightarrow CO_2(^1\Sigma)$$

where κ may be 10^{-5} or lower. The universal validity of the principle has been questioned and it should certainly be used with great caution.

Symmetry and statistical factor

If molecules involved in a reaction have symmetry some sophisticated complications arise in the choice of partition functions for the species involved and so in calculating the rate equation. Where molecules possess rotational symmetry, the corresponding partition functions should be divided by the appropriate symmetry numbers when calculating conventional equilibrium constants. However, where the equilibrium constants for formation of the activated complex are concerned, difficulties can arise which are best resolved by omitting symmetry numbers from the partition functions and, instead, multiplying the result by a statistical factor. This statistical factor is the number of different activated complexes that can be formed if all the identical atoms are labelled to distinguish them. Great care must be exercised in the choice of activated complex with the correct degree of symmetry. This oversimplified discussion can be misleading and the student is referred to more comprehensive treatments[15].

Quantum mechanical tunnelling

In classical mechanics the only way to pass the potential energy barrier is over the top. With quantum mechanics there is a finite probability, however small, that a system with less energy than is required to pass over the barrier can still appear on the other side. This phenomenon is called *tunnelling*. It is most likely when the energy barrier is low and thin, and when the reacting particle is small. For example it is more likely with hydrogen atoms than with the deuterium isotope, and this should be taken into account when discussing hydrogen–deuterium isotope effects. The lighter isotope may be found to react more rapidly than can be explained by other factors. The quantitative treatment of tunnelling in reaction kinetics is very difficult, and in many aspects is still unclear, and again the student is referred to more advanced texts[15].

5.6 Comparison of ACT Results with Experiment

Although the possible complications discussed above should be borne in mind, this section is based on the simple result of the ACT, with transmission coefficient unity:

$$k = \frac{\bar{k}T}{h}\frac{Q_{\ddagger}'}{Q_A Q_B}e^{-E_0^{\ddagger}/RT} \qquad (5.5.9)$$

When compared with the Arrhenius equation this result is seen to have an exponential term of correct form. The pre-exponential factor requires detailed examination. It has a complicated dependence on temperature, since all partition functions have temperature

[15] K. J. Laidler, *Theories of Chemical Reaction Rates*, McGraw-Hill, 1969.

terms, but cancellation of similar factors may leave only small residual temperature variation. The designation 'universal frequency factor' has sometimes been applied to $\bar{k}T/h$. Near room temperature its value is about $6 \times 10^{12}\ \mathrm{s}^{-1}$, close to typical vibrational frequencies. The partition function term reduces to units of (volume)$^{-1}$ when the expressions in table 5.2 are used, and conversion to any concentration units is quite straightforward.

The pre-exponential factor will now be calculated for the reaction of two hard spheres, and this will be followed by an approximate account of more realistic cases.

For reaction between hard spheres:

$$A + B \rightleftharpoons \ddagger \rightarrow \text{Products}$$

the reactants have only translational degrees of freedom and $Q_A = q_{t(A)}^3$ and $Q_B = q_{t(B)}^3$. The activated complex is a diatomic species with three translational and two rotational degrees of freedom. In a normal diatomic there is one vibration, but in activated complexes it has been seen, in the previous section, that one vibrational term is missing in $Q_\ddagger'$. Thus $Q_\ddagger' = q_{t(\ddagger)}^3\, q_{r(\ddagger)}^2$ in this case.

Hence, from (5.5.9)

$$k = \frac{\bar{k}T}{h} \frac{q_{t(\ddagger)}^3\, q_{r(\ddagger)}^2}{q_{t(A)}^3 q_{t(B)}^3}\, \mathrm{e}^{-E_0\ddagger/RT} \qquad (5.6.1)$$

The masses of reactants and complex are m_A, m_B and $(m_A + m_B)$ respectively, the atomic radii are r_A, r_B, and if $(r_A + r_B)$ is taken for the radius of the complex, the appropriate expressions from table 5.2 may be substituted into (5.6.1), giving:

$$k = \pi(r_A + r_B)^2 \left(\frac{8\bar{k}T}{\pi\mu}\right)^{1/2} \mathrm{e}^{-E_0\ddagger/RT}$$

This is equivalent to the SCT result, equation (5.1.11) derived in section 5.1. For more realistic cases, full calculations are complicated but there is a convenient approximate approach. This is to assume that for reactants and activated complex all translational partition functions are equal ($q_t \sim 10^8\ \mathrm{cm}^{-1}$ for each degree of freedom). Similarly all rotational and vibrational partition functions are assumed equal ($q_r \sim 10$ and $q_v \sim 1$ for each degree of freedom). With these approximations some important features emerge clearly.

For reactants, where N is the number of atoms in a polyatomic species, the total partition functions are:

Atoms	$Q = q_t^{\,3}$
Linear polyatomics	$Q = q_t^{\,3} q_r^{\,2} q_v^{\,3N-5}$
Non-linear polyatomics	$Q = q_t^{\,3} q_r^{\,3} q_v^{\,3N-6}$

The activated complex is represented by the appropriate linear or non-linear expression with *one fewer* vibrational term, as explained in the previous section. Consider some common reaction types of increasing complexity.

(i) SCT hard-sphere approximation:

Atom + atom $\rightleftharpoons$ diatomic activated complex

Substitution into (5.5.9) gives:

$$k \sim \frac{\bar{k}T}{h}\frac{q_t^3 q_r^2}{q_t^3 \times q_t^3} e^{-E_0^{\ddagger}/RT}$$

and, as stated above, assuming that all translational partition functions are the same:

$$k \sim \frac{\bar{k}T}{h}\left(\frac{q_r^2}{q_t^3}\right) e^{-E_0^{\ddagger}/RT}$$

(ii) atom + linear molecule $\rightleftharpoons$ linear complex

Substitution into (5.5.9) gives:

$$k \sim \frac{\bar{k}T}{h}\frac{q_t^3 q_r^2 q_v^{3(N+1)-6}}{q_t^3 \times q_t^3 q_r^2 q_v^{3N-5}} e^{-E_0^{\ddagger}/RT}$$

yielding:

$$k \sim \frac{\bar{k}T}{h}\left(\frac{q_v^2}{q_t^3}\right) e^{-E_0^{\ddagger}/RT}$$

The same result is obtained for:

Atom + non-linear molecule $\rightleftharpoons$ non-linear activated complex

(iii) linear molecule + linear molecule $\rightleftharpoons$ linear complex

The same approach gives:

$$k \sim \frac{\bar{k}T}{h}\left(\frac{q_v^4}{q_r^2 q_t^3}\right) e^{-E_0^{\ddagger}/RT}$$

which is also the result obtained for:

Linear molecule + non-lincar molecule $\rightleftharpoons$ non-linear complex

(iv) non-linear molecule + non-linear molecule $\rightleftharpoons$ non-linear complex

The method gives:

$$k \sim \frac{\bar{k}T}{h}\left(\frac{q_v^5}{q_r^3 q_t^3}\right) e^{-E_0^{\ddagger}/RT}$$

It is a simple matter to make order of magnitude estimates for the value of the bimolecular rate constants in each case using $q_t \sim 10^8$ cm^{-1}, $q_r \sim 10$, $q_v \sim 1$ and $\bar{k}T/h \sim 10^{13}$ s^{-1}. This leads to rate constants in units of molecule^{-1} cm^3 s^{-1} which may be transformed into molar units by multiplying with the Avogadro constant. The results for cases (i) to (iv) respectively are about 6×10^{14}, 6×10^{12}, 6×10^{10}, 6×10^9 mol^{-1} cm^3 s^{-1}. Hence, as the molecular complexity of the reaction increases, k should progressively decrease. This general trend has already been noted for the reactions listed in table 5.1 and is further illustrated in table 5.3. In this respect therefore, the ACT represents a clear advance on the SCT.

The exact calculation of pre-exponential factors requires exact values for the partition functions of reactants and activated complex. For many reactant species accurate

Table 5.3
Pre-exponential factors for some bimolecular reactions calculated by ACT compared with experimental values.

Reaction	A_{exp}/ $mol^{-1}\ cm^3\ s^{-1}$	A_{calc}/ $mol^{-1}\ cm^3\ s^{-1}$
$H + H_2 \rightarrow H_2 + H$	$5{\cdot}4 \times 10^{13}$	$7{\cdot}4 \times 10^{13}$
$Br + H_2 \rightarrow HBr + H$	3×10^{13}	1×10^{14}
$H + CH_4 \rightarrow H_2 + CH_3$	1×10^{13}	2×10^{13}
$H + C_2H_6 \rightarrow H_2 + C_2H_5$	3×10^{12}	1×10^{13}
$CH_3 + H_2 \rightarrow CH_4 + H$	2×10^{12}	1×10^{12}
$CH_3 + CH_3COCH_3 \rightarrow CH_4 + CH_3COCH_2$	4×10^{11}	1×10^{11}
$CD_3 + CH_4 \rightarrow CD_3H + CH_3$	1×10^{11}	2×10^{11}
$2ClO \rightarrow Cl_2 + O_2$	6×10^{10}	1×10^{11}
$F_2 + ClO_2 \rightarrow FClO_2 + F$	3×10^{10}	8×10^{10}
$CH_3 + isoC_4H_{10} \rightarrow CH_4 + C_4H_9$	1×10^{10}	6×10^{9}

spectroscopic measurements of bond lengths and vibrational frequencies allow this calculation, but frequencies and geometrical parameters must be *estimated* for the activated complex. The choice of reaction coordinate is very important and can only be made reliably for a known potential energy surface. Since construction of surfaces is so difficult this is a great obstacle to exact calculations. The difficulties are such that often experimental results are used to deduce properties of the activated complex. Such knowledge can be useful for a semi-empirical approach to a widening range of reactions.

Despite these limitations the ACT has been successfully applied to many reactions as illustrated by the calculated values quoted in table 5.3.

Activation energy and $E_0^{\ddagger}$

The temperature variation exhibited by the ACT pre-exponential factor is complicated since the partition function for each degree of freedom varies with temperature. The total effect depends on the molecular detail of each specific reaction. Inspection of the ACT equations for the rate constant shows that they may be expressed:

$$k = CT^n e^{-E_0^{\ddagger}/RT} \qquad (5.6.2)$$

where C is the temperature-independent term and n the total temperature exponent. For the cases (i)–(iv) discussed above, taking the temperature dependence of q_v as approximately T^0, the n values are $\frac{1}{2}$, $-\frac{1}{2}$, $-\frac{3}{2}$ and -2 respectively. In each case the variation represented by this term is small compared to the exponential term.

The formal relationship between experimental activation energy, and $E_0^{\ddagger}$ is easily obtained:

$$E_{exp} = RT^2\left(\frac{d(\ln k)}{dT}\right) \qquad (1.10.2)$$

and substituting (5.6.2) for the variation of k with temperature gives

$$E_{exp} = E_0^{\ddagger} + nRT \qquad (5.6.3)$$

Though this relationship is quite explicit it must be remembered that for most reactions, calculation of potential energy surfaces and hence $E_0^{\ddagger}$ is beyond the theoreticians' scope at present.

5.7 Thermodynamic Formulation of ACT

Partition functions for species in solution are extremely difficult or even impossible to calculate and there is some difficulty for complex molecules in the gas phase. In such cases the theory is more useful when reformulated in thermodynamic terms.

The rate constant expression (5.5.9):

$$k = \frac{\bar{k}T}{h} \frac{Q_{\ddagger}'}{Q_A Q_B} e^{-E_0^{\ddagger}/RT}$$

can be written

$$k = \frac{\bar{k}T}{h} K_c^{\ddagger} \qquad (5.7.1)$$

where

$$K_c^{\ddagger} = \frac{Q_{\ddagger}'}{Q_A Q_B} e^{-E_0^{\ddagger}/RT} \qquad (5.7.2)$$

The right-hand side of (5.7.2) is, formally, a statistical mechanics expression for an equilibrium constant, though it involves a molecule with a degree of vibrational freedom missing. The subscript for $K_c^{\ddagger}$ emphasizes that it is a concentration constant and:

$$-RT \ln K_c^{\ddagger} = \Delta G_0^{\ddagger} \qquad (5.7.3)$$

where $\Delta G_0^{\ddagger}$ is the standard Gibbs free energy change for the process, reactants →activated complex, the *free energy of activation* (standard states being defined in terms of concentrations†). Substitution into (5.7.1) gives

$$k = \frac{\bar{k}T}{h} e^{-\Delta G_0^{\ddagger}/RT} \qquad (5.7.4)$$

From the thermodynamic definition of free energy change:

$$\Delta G_0^{\ddagger} = \Delta H_0^{\ddagger} - T\Delta S_0^{\ddagger}$$

† Thermodynamics leads to both

$$\Delta G = -RT \ln K_p + RT \sum_i n_i \ln p_i$$

and

$$\Delta G = -RT \ln K_c + RT \sum_i n_i \ln c_i \qquad (5.7.3a)$$

where p_i and c_i are the pressures and concentrations of gaseous species and n_i the number of moles of each in the stoichiometric equation. K_p and K_c are the equilibrium constants expressed in terms of pressures and concentrations respectively. In conventional thermodynamics the standard state is a pressure of one atmosphere (760 Torr, 101 325 N m^{-2}) and pressures are expressed in atmospheres. This leads to the standard Gibbs free energy change given by $\Delta G_0 = -RT \ln K_p$. K_p is related to K_c by $K_p = K_c(RT)^{\Delta n}$ where Δn is the change in number of moles going from initial to final state in the stoichiometric equation. Taking unit concentration as the standard state leads from equation (5.7.3a) to the standard Gibbs free energy change defined by equation (5.7.3) above as used in the conventional kinetic derivation. It must be remembered however that standard compilations of thermodynamic data refer to the standard state of unit pressure.

where $\Delta H_0^{\ddagger}$ and $\Delta S_0^{\ddagger}$ are the standard enthalpy and entropy changes, reactant → activated complex, the *enthalpy* and *entropy of activation.* The standard state for these thermodynamic functions is also unit concentration, in the same units in which $K_c^{\ddagger}$ is evaluated. Substitution into (5.7.4) gives:

$$k = \frac{\bar{k}T}{h} e^{\Delta S_0^{\ddagger}/R} e^{-\Delta H_0^{\ddagger}/RT} \tag{5.7.5}$$

This has the Arrhenius form, with $\Delta H_0^{\ddagger}$ approximating to an activation energy term. More importantly a whole new interpretative world is opened by identifying the experimental pre-exponential factor with a function of the entropy of activation. This has provided many useful insights particularly in solution kinetics.

Experimental activation energy and $\Delta H_0^{\ddagger}$

The activation energy is given by

$$E_{exp} = RT^2\left(\frac{d(\ln k)}{dT}\right)$$

and substituting equation (5.7.1) yields

$$E_{exp} = RT^2\left\{\frac{d}{dT}\ln\left(\frac{\bar{k}T}{h}K_c^{\ddagger}\right)\right\} = RT + RT^2\left(\frac{d(\ln K_c^{\ddagger})}{dT}\right)$$

The temperature variation of a concentration equilibrium constant is:

$$\frac{d(\ln K)}{dT} = \frac{\Delta E_0}{RT^2}$$

and so

$$E_{exp} = RT + \Delta E_0^{\ddagger}$$

where $\Delta E_0^{\ddagger}$ is the standard energy change, reactants → activated complex. The thermodynamic definition of enthalpy

$$\Delta H_0^{\ddagger} = \Delta E_0^{\ddagger} + \Delta(PV)^{\ddagger}$$

gives

$$E_{exp} = \Delta H_0^{\ddagger} + RT - \Delta(PV)^{\ddagger} \tag{5.7.6}$$

For solution reactions, where volume changes are very small, at approximately constant pressure $\Delta(PV)^{\ddagger} \sim 0$ and:

$$E_{exp} = \Delta H_0^{\ddagger} + RT \tag{5.7.7}$$

Substituting (5.7.7) into equation (5.7.5) yields

$$k = e\frac{\bar{k}T}{h} e^{\Delta S_0^{\ddagger}/R} e^{-E_{exp}/RT} \tag{5.7.8}$$

For gas reactions, at normal temperatures and pressures where behaviour is approximately ideal:

$$\Delta(PV)^{\ddagger} = \Delta n^{\ddagger}RT \tag{5.7.9}$$

where $\Delta n^{\ddagger}$ is the change in number of moles going from reactants → activated complex.

From (5.7.6) and (5.7.9)

$$E_{exp} = \Delta H_0^{\ddagger} - (\Delta n^{\ddagger} - 1)RT \qquad (5.7.10)$$

and substituting into equation (5.7.5) gives

$$k = e^{-(\Delta n^{\ddagger} - 1)} \frac{\bar{k}T}{h} e^{\Delta S^{\ddagger}/R} e^{-E_{exp}/RT} \qquad (5.7.11)$$

For example, in the bimolecular case $\Delta n^{\ddagger} = -1$ and

$$k = e^2 \frac{\bar{k}T}{h} e^{\Delta S^{\ddagger}/R} e^{-E_{exp}/RT}$$

Free energy surfaces

The thermodynamic formulation in equation (5.7.4) relates the rate constant to free energy of activation and universal constants only. It has been suggested that reactants and activated complex might be better represented on a free energy surface, rather than a potential energy surface. For the motion of individual species from reactant to product configuration, potential energy is the determining factor. With an assembly of molecules in a statistical distribution among energy states free energy determines whether spontaneous motion occurs in a given direction. Such a surface would vary with temperature and is identical with the potential energy surface only at absolute zero of temperature. However, no free energy surface appears to have been constructed and it presents no easy task.

5.8 Termolecular Reactions

There are two main categories of such reactions:

(i) A small group of reactions for nitric oxide:

$2NO + O_2 \rightarrow 2NO_2$

$2NO + Cl_2 \rightarrow 2NOCl$

$2NO + Br_2 \rightarrow 2NOBr$

These have pre-exponential factors of about 10^9–10^{10} mol^{-2} cm^6 s^{-1}, and activation energies close to zero or even with small negative values.

(ii) The combination of atoms or small radicals in the presence of a third body, in general:

$A + A + M \rightarrow A_2 + M$

$A + B + M \rightarrow AB + M$

These have higher pre-exponential factors of about 10^{14}–10^{16} mol^{-2} cm^6 s^{-1} and again activation energies are negative. In addition, for a given atom combination, rate constants for different third bodies vary widely—there is a wide range of efficiencies for third bodies, differing by factors of up to 10^3. A good example is provided by the combination of iodine atoms, as shown in table 5.4.

Table 5.4
Rate constants and Arrhenius parameters for the reaction $I + I + M \rightarrow I_2 + M$. (From G. Porter and J. A. Smith, *Proc. R. Soc.* (*London*), **A261**, 28 (1961).)

Third body (M)	k_{exp} (293 K)/ 10^{16} mol^{-2} cm^6 s^{-1}	A_{exp}/ 10^{15} mol^{-2} cm^6 s^{-1}	E_{exp}/ kJ mol^{-1}
helium	0·3	1·5	1·7
argon	0·59	0·61	5·4
oxygen	1·3	1·0	6·3
carbon dioxide	2·7	1·3	7·3
benzene	16	8·6	7·1
toluene	39	3·7	11
ethyliodide	52	8·1	10
mesitylene	81	0·69	17
iodine	274	0·65	21

Large polyatomic radicals undergo bimolecular combination, and methyl radicals show intermediate behaviour, the kinetics being second order at high pressure, but third order at very low pressure. This is analogous to the behaviour of the reverse process, dissociation, in varying from first order at high pressure to second order at low pressure, as discussed for unimolecular reactions in the next sections.

SCT for termolecular reactions

The three-body collision rate can be calculated by assuming that two molecules must be within a certain distance (their joint collision diameter) when struck by a third. Such calculations give pre-exponential factors in the range 10^{14}–10^{16} mol^{-2} cm^6 s^{-1}, but there is no way of predicting different third body efficiencies or negative activation energies.

ACT for termolecular reactions

The model is:

$$A + B + C \rightleftharpoons ABC^{\ddagger} \rightarrow \text{products}$$

A derivation, similar to that given for bimolecular reactions, yields:

$$k = \frac{\bar{k}T}{h} \frac{Q_{\ddagger}'}{Q_A Q_B Q_C} \mathrm{e}^{-E_0{\ddagger}/RT} \tag{5.8.1}$$

and detailed calculations on this basis, with suitable assumptions for the form of the activated complex, give quite good agreement for the pre-exponential factor.

Equation (5.8.1) may be expressed

$$k = CT^n \mathrm{e}^{-E_0{\ddagger}/RT}$$

where C is the temperature-independent part of the pre-exponential factor and exponent n is obtained from the temperature dependence of the molecular partition functions. Since:

$$E_{exp} = RT^2 \left(\frac{\mathrm{d}(\ln k)}{\mathrm{d}T}\right) \tag{5.8.2}$$

$$E_{exp} = E_0^{\ddagger} + nRT$$

For termolecular reactions $n = -3$ is typical and if $E_0^{\ddagger}$ is very small this gives a negative activation energy. This success cannot be extended to explain the range of observed third body efficiencies. A new insight is required for the atom combination reactions.

Atom combination reactions

The problems are substantially resolved by postulating a mechanism of consecutive bimolecular steps for the so-called termolecular processes[16]. Two possibilities are:

(i) The energy transfer mechanism:

$$A + A \rightarrow A_2^*$$

$$A_2^* \rightarrow A + A$$

$$A_2^* + M \rightarrow A_2 + M$$

(ii) The intermediate complex mechanism:

$$A + M \xrightarrow{k_1} A..M \qquad (5.8.\text{I})$$

$$A..M \xrightarrow{k_{-1}} A + M \qquad (5.8.\text{II})$$

$$A..M + A \xrightarrow{k_2} A_2 + M \qquad (5.8.\text{III})$$

The transient species A_2^* and $A..M$ are true molecular entities *not* activated complexes. The energy transfer mechanism is more successful for some reactions, the intermediate complex for others and frequently both must contribute. A notable example is the combination of iodine atoms and mechanism (ii) will be chosen to demonstrate that it is consistent with third-order kinetics, varying third body efficiencies and negative activation energy.

Interaction between the atom and third body is relatively weak, intermediate $A..M$ will be short lived, and the stationary-state hypothesis may be applied:

$$\mathrm{d}[A..M]/\mathrm{d}t = k_1[A][M] - k_{-1}[A..M] - k_2[A..M][A] = 0$$

$$[A..M] = k_1[A][M]/(k_{-1} + k_2[A])$$

The rate of product formation is

$$\mathrm{d}[A_2]/\mathrm{d}t = k_2[A..M][A] = k_1k_2[A]^2[M]/(k_{-1} + k_2[A])$$

For a weak complex, more likely to dissociate than survive until reaction, $k_{-1} \gg k_2[A]$ and

$$\frac{\mathrm{d}[A_2]}{\mathrm{d}t} = \frac{k_1k_2}{k_{-1}}[A]^2[M]$$

The equilbrium constant for formation of $[A..M]$ is $K = k_1/k_{-1}$ thus

$$\mathrm{d}[A_2]/\mathrm{d}t = Kk_2[A]^2[M]$$

[16] G. Porter, *Disc. Faraday Soc.*, **33**, 198 (1962).

Hence:

(a) The mechanism gives third-order kinetics with experimental rate constant $k_{exp} = Kk_2$.

(b) The stronger the $A..M$ interaction the greater the value of K, and so the higher the experimental rate constant. With iodine atoms this effect is enhanced if M is an aromatic hydrocarbon which can form a charge-transfer complex, and especially so where M is the iodine molecule ($I + I_2 \rightleftharpoons I_3$). Calculations for K and the experimental rate constant have given good agreement with the range of third body efficiencies found for iodine combination.

(c) The experimental activation energy is given:

$$E_{exp} = RT^2\left(\frac{d(\ln k)}{dT}\right) = RT^2\left(\frac{d(\ln Kk_2)}{dT}\right)$$

$$= RT^2\left(\frac{d(\ln K)}{dT}\right) + RT^2\left(\frac{d(\ln k_2)}{dT}\right)$$

The first term on the right-hand side, the temperature variation of the *equilibrium* constant, is given by $\Delta E_1°$, the standard energy change for reaction 1. The second term is simply the temperature variation of the *rate* constant, given by E_2 the activation energy for reaction 2. Thus:

$$E_{exp} = \Delta E_1° + E_2 \qquad (5.8.3)$$

$\Delta E_1°$ is the energy released on forming the weak bond in the complex. It will be a small negative term, varying with the strength of the complex. In reaction 2, a weak bond is broken and a strong bond formed giving high exothermicity and the activation energy E_2 should be very small. Thus, from equation (5.8.3), the experimental activation should be negative, increasingly so as the strength of the complex increases.

This account still represents a considerable simplification for termolecular combinations and the reverse dissociation of diatomic molecules, where the measured activation energy is often less than the bond dissociation energy. The reverse reactions must involve highly internally excited states of the diatomics. These internal energy states are not in thermal equilibrium populations and a multi-step approach allowing for energy exchange among many energy levels (network effects[17]) must be adopted.

5.9 Unimolecular Reactions, Lindemann's Mechanism

Some large groups of reactions such as dissociation or isomerization obey first-order kinetics, and some examples are given in table 5.5. For the overall reaction: $A \rightarrow$ products, the experimental rate is:

$$d[prod.]/dt = k_{exp}[A]$$

[17] H. O. Pritchard, *Specialist Periodical Report*: *Reaction Kinetics*, Vol. 1, Chemical Society London, 1975 p. 243.

Table 5.5
Experimental Arrhenius parameters for unimolecular reactions. The pre-exponential factor is presented in logarithmic form, e.g. for the last reaction quoted $A = 10^{13}\ s^{-1}$.

Reaction	$\log_{10}(A_{exp}/s^{-1})$	$E_{exp}/kJ\ mol^{-1}$
Molecular isomerization		
*cis*1,2 dideuterocyclo propane → *trans*	16·4	272
cyclo propane → propylene	15·2	272
$CH_3NC \rightarrow CH_3CN$	13·6	160
*cis*CHD=CHD → *trans*	12·5	256
Molecular dissociation to molecules		
cyclobutane → $2C_2H_4$	15·6	261
$C_2H_5I \rightarrow C_2H_4 + HI$	13·4	209
$(COOH)_2 \rightarrow HCOOH + CO_2$	11·9	125
Molecular dissociation to radicals		
$C_2H_6 \rightarrow 2CH_3$	17·4	384
$(CH_3)_3COCO(CH_3)_3 \rightarrow 2(CH_3)_3CO$	15·6	156
$N_2O_5 \rightarrow NO_2 + NO_3$	14·8	88
$N_2O \rightarrow N_2 + O$	11·9	250
Radical dissociation		
$CH_3CO \rightarrow CH_3 + CO$	15	43
$nC_3H_7 \rightarrow C_3H_6 + H$	14·6	158
$C_2H_5 \rightarrow C_2H_4 + H$	13	167

The main feature of such reactions is: as initial reactant pressure decreases k_{exp} falls, and at very low pressures the reaction is second order. This behaviour is shown schematically in figure 5.12. The limiting high-pressure value of the first-order constant is designated k_{exp}^{∞}. The diagram shows a 'fall-off' region which is conveniently characterized for a particular reaction by the pressure, or concentration $[A]_{1/2}$, at which the rate constant has fallen to $k_{exp}^{\infty}/2$. A quantitative interpretation of these features will tax any kinetic theory and many have been advanced, starting with what are essentially collision theories.

All theoretical approaches are based on an insight by Lindemann[18]. He realized that a unimolecular reaction was not a single elementary step, but must involve a mechanism in which some molecules are highly energized by energy transfer on collision and only these energized molecules undergo true elementary unimolecular reaction to give products.

Lindemann's mechanism is

$$A + A \xrightarrow{k_1} A^* + A \quad (5.9.I)$$

$$A + A^* \xrightarrow{k_{-1}} A + A \quad (5.9.II)$$

$$A^* \xrightarrow{k_2} \text{products} \quad (5.9.III)$$

[18] F. A. Lindemann, *Trans. Faraday Soc.*, **17**, 598 (1922). Most kineticists have lost sight of the fact that the solution was proposed independently and almost simultaneously by J. A. Christiansen, *Reaktionskinetiske Studier*, Copenhagen, 1921.

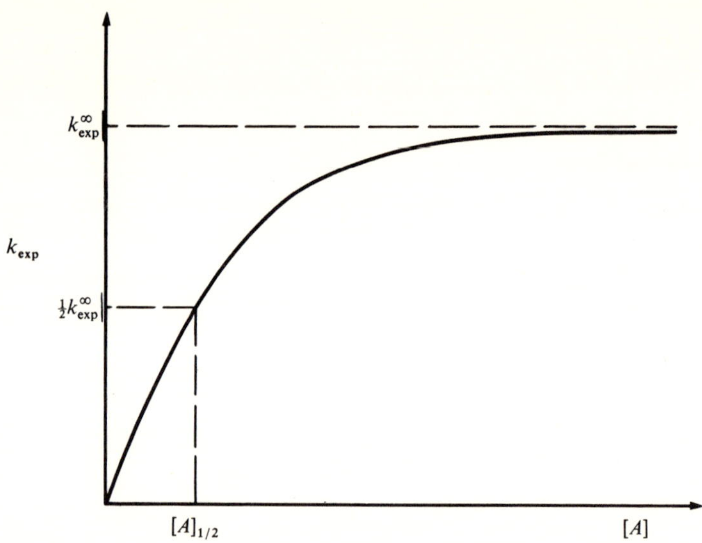

Figure 5.12
The variation of an experimental unimolecular rate constant with pressure, shown as the equivalent variation with concentration of reactant, $[A]$.

The energized molecule formed in step (5.9.I) may be de-energized in collisions (5.9.II) or react in the competing unimolecular step, (5.9.III). The energized molecule, which is *not* an activated complex, will have a very short lifetime, and applying the stationary-state hypothesis:

$$d[A^*]/dt = k_1[A]^2 - k_{-1}[A^*][A] - k_2[A^*] = 0$$

Hence

$$[A^*] = k_1[A]^2/(k_{-1}[A] + k_2)$$

The rate of reaction may be expressed

$$d[\text{prod.}]/dt = k_2[A^*]$$

Hence

$$d[\text{prod.}]/dt = k_1k_2[A]^2/(k_{-1}[A] + k_2) \tag{5.9.1}$$

This rate equation covers the entire concentration or pressure range. At *high pressure* $k_{-1}[A] > k_2$, collisional de-activation of A^* is more probable than reaction. This gives, from (5.9.1):

$$d[\text{prod.}]/dt = (k_1k_2/k_{-1})[A] \tag{5.9.2}$$

in agreement with the observed first-order kinetics at high pressure. The experimental rate equation at the high-pressure limit is:

$$d[\text{prod.}]/dt = k_{exp}^{\infty}[A]$$

and comparison with (5.9.2) shows:

$$k^{\infty}_{\text{exp}} = k_1 k_2 / k_{-1} \qquad (5.9.3)$$

At very *low pressure* $k_{-1}[A] < k_2$, collisional de-activation is slower than reaction, and from (5.9.1):

$$\text{d[prod.]}/\text{d}t = k_1[A]^2$$

giving the observed second-order kinetics. The rate over the *complete pressure range* may be expressed in an alternative way. The experimental rate over the entire range is:

$$\text{d[prod.]}/\text{d}t = k_{\text{exp}}[A]$$

The theoretical rate over the whole range is given by equation (5.9.1) and comparing terms shows:

$$k_{\text{exp}} = \frac{k_1 k_2[A]}{k_{-1}[A] + k_2} = \frac{k_1 k_2 / k_{-1}}{1 + k_2/k_{-1}[A]} \qquad (5.9.4)$$

Substitution of (5.9.3) in (5.9.4) gives:

$$k_{\text{exp}} = \frac{k^{\infty}_{\text{exp}}}{1 + k_2/k_{-1}[A]} \qquad (5.9.5)$$

A plot of k_{exp} versus $[A]$ has the same qualitative form as shown in figure 5.12, and so the theory is satisfactory in this respect.

The *fall-off region* is conveniently characterized by the value of $[A]_{1/2}$, the concentration where $k_{\text{exp}} = k^{\infty}_{\text{exp}}/2$. From (5.9.5), $[A]_{1/2} = k_2/k_{-1}$, and using (5.9.3), $[A]_{1/2} = k^{\infty}_{\text{exp}}/k_1$. Since k^{∞}_{exp} and $[A]_{1/2}$ are easily measured experimentally this provides a convenient measurement of k_1, the constant for an energy transfer process. Conversely, calculation of k_1 allows prediction of the fall-off region for the reaction.

The Lindemann theory is qualitatively consistent with experimental data, and it may be tested quantitatively:

(i) Measured values of k_1 may be compared with theoretical calculations for this bimolecular process. k_1 turns out to be several orders of magnitude *greater* than allowed by the SCT, a deviation opposite to that found with other bimolecular reactions.

(ii) Equations (5.9.4) and (5.9.3) can be re-arranged to

$$\frac{1}{k_{\text{exp}}} = \frac{1}{k_1[A]} + \frac{1}{k^{\infty}_{\text{exp}}} \qquad (5.9.6)$$

Whatever the value of k_1, a plot of $1/k_{\text{exp}}$ against $1/[A]$ should give a straight line. In practice, marked deviations from linearity are found.

Since Lindemann's pioneering work improvements in unimolecular theory have essentially been aimed at re-interpreting k_1 and k_2, leading to a better result for the variation of k_{exp} with $[A]$.

5.10 Hinshelwood's Modification

The SCT calculation for k_1 is based on energy transfer along the line of centres when molecules collide. Hinshelwood[19], in 1927, took a more realistic model which includes vibrational modes for the molecules exchanging energy.

In section 5.2, it was seen that the fraction of molecules with energy between ε and $\varepsilon + \mathrm{d}\varepsilon$, $F(\varepsilon)\,\mathrm{d}\varepsilon$, was given by equation (5.2.1). Molecules with this energy make a contribution $\mathrm{d}k_1$ to the rate constant k_1, and the total value of k_1 is the sum of such contributions for all possible energies. The energized molecules are very efficiently removed in collisions, the rate parameter for this process being k_{-1}. The above fraction $F(\varepsilon)\,\mathrm{d}\varepsilon$ is, in fact, equal to $\mathrm{d}k_1/k_{-1}$ and from equation (5.2.1):

$$\frac{\mathrm{d}k_1}{k_{-1}} = \frac{1}{(n-1)!}\left(\frac{\varepsilon}{\bar{k}T}\right)^{n-1} \frac{\mathrm{e}^{-\varepsilon/\bar{k}T}}{\bar{k}T}\,\mathrm{d}\varepsilon \qquad (5.10.1)$$

k_1/k_{-1} is thus obtained by integrating (5.10.1) from ε^* (the minimum energy needed for the unimolecular step) to infinity. If $\varepsilon^* \gg (n-1)\bar{k}T$ this gives:

$$\frac{k_1}{k_{-1}} = \frac{1}{(n-1)!}\left(\frac{\varepsilon^*}{\bar{k}T}\right)^{n-1} \mathrm{e}^{-\varepsilon^*/\bar{k}T}$$

The de-energization of A^* on collision is, as stated, very efficient and $k_{-1} \simeq Z_{-1}$ the collision frequency. Hence:

$$k_1 \simeq Z_{-1}\frac{1}{(n-1)!}\left(\frac{\varepsilon^*}{\bar{k}T}\right)^{n-1} \mathrm{e}^{-\varepsilon^*/\bar{k}T} \qquad (5.10.2)$$

If (5.10.2) is compared with the SCT result (5.1.10) it is seen that the rate constant for the energy transfer process exceeds the SCT constant by the factor

$$\frac{1}{(n-1)!}\left(\frac{\varepsilon^*}{\bar{k}T}\right)^{n-1}$$

For reasonable values of n and ε^*, this factor may be 10^5 or greater, thus giving a good account of the high values of k_1 found experimentally.

However, there remain other inconsistencies between experiment and the Lindemann–Hinshelwood treatment. For example, irrespective of the value of k_1, the theory still fails to give quantitative agreement for the variation of k_{exp} with $[A]$ in the fall-off and very low-pressure regions. Further improvement requires a more detailed examination of k_2.

5.11 Rice, Ramsperger, Kassel (RRK) Statistical Treatment

In 1927–8 Rice and Ramsperger, and Kassel[20] independently treated the energy distribution in normal modes of a vibrating molecule on a statistical basis, with k_2 expressed as a function of the energy of the energized molecule, A^*. Molecules are

[19] C. N. Hinshelwood, *Proc. R. Soc. (London)*, **A113**, 230 (1927).
[20] O. K. Rice and H. C. Ramsperger, *J. Am. Chem. Soc.*, **49**, 1616 (1927); *J. Am. Chem. Soc.*, **50**, 617 (1928). L. S. Kassel, *J. Phys. Chem.*, **32**, 225 (1928).

assumed to consist of loosely-coupled oscillators, allowing energy flow between vibrational modes.

For molecules having energy in the range $\varepsilon \to \varepsilon + d\varepsilon$ the statistical probability that an amount ε^* is in the critical mode leading to reaction is given:

$$\text{Probability} = \left(\frac{\varepsilon - \varepsilon^*}{\varepsilon}\right)^{n-1} \tag{5.11.1}$$

where n is the number of modes. When ε^* is present in the critical mode reaction follows, and the rate at which energy passes into the mode is proportional to the above probability. This model also suggests that the proportionality constant should be of the order of a vibrational frequency, ν. Thus:

$$k_2 = \nu\left(\frac{\varepsilon - \varepsilon^*}{\varepsilon}\right)^{n-1} \tag{5.11.2}$$

The experimental rate constant at high pressure may be deduced from equation (5.9.3):

$$k_{\text{exp}}^{\infty} = k_2\left(\frac{k_1}{k_{-1}}\right) = \int_{\varepsilon^*}^{\infty} k_2\left(\frac{dk_1}{k_{-1}}\right)$$

Taking equation (5.10.1) for dk_1/k_{-1} and (5.11.2) for k_2 leads to

$$k_{\text{exp}}^{\infty} = \int_{\varepsilon^*}^{\infty} \nu\left(\frac{\varepsilon - \varepsilon^*}{\varepsilon}\right)^{n-1} \frac{1}{(n-1)!}\left(\frac{\varepsilon}{\bar{k}T}\right)^{n-1} \frac{e^{-\varepsilon/\bar{k}T}}{\bar{k}T}\, d\varepsilon$$

which reduces to:

$$k_{\text{exp}}^{\infty} = \nu\, e^{-\varepsilon^*/\bar{k}T} \tag{5.11.3}$$

That ν is of the order of a vibrational frequency in many cases is confirmed by comparison with the experimental pre-exponential factors quoted in table 5.5.

The pressure variation, k_{exp} versus $[A]$, is obtained in an analogous manner. Equation (5.9.4) gives:

$$dk_{\text{exp}} = \frac{k_2(dk_1/k_{-1})}{1 + k_2/k_{-1}[A]} \tag{5.11.4}$$

Insertion of values for k_2 and (dk_1/k_{-1}) and integration from ε^* to infinity gives, for the variation of k_{exp} with $[A]$, the very complicated function:

$$k_{\text{exp}} = \frac{\nu\, e^{-\varepsilon^*/\bar{k}T}}{(n-1)!} \int_0^{\infty} \frac{x^{n-1}\, e^{-x}\, dx}{1 + (\nu/k_{-1}[A])(x/[x+b])^{n-1}} \tag{5.11.5}$$

where $x = (\varepsilon - \varepsilon^*)/\bar{k}T$ and $b = \varepsilon^*/\bar{k}T$. k_{exp} can be calculated for various values of $[A]$, but n must be *chosen* to give the best fit to experimental data. Satisfactory agreement is usually obtained for n approximately half the number of normal modes in the molecule, but the correct n value cannot be predicted. Similarly, although ν is often found experimentally to be about $10^{13}\ s^{-1}$, the order of magnitude of vibrational frequencies, it can be several orders of magnitude higher as shown in table 5.5.

5.12 ACT for Unimolecular Reactions

Following the historical sequence of unimolecular theories, in the 1930's the ACT was applied to unimolecular rate constants for the high-pressure region (k_{exp}^{∞}). A Boltzmann energy distribution cannot be assumed for the low-pressure regions.

The model is

$$A \rightleftharpoons A^{\ddagger} \rightarrow \text{products}$$

irrespective of the detailed mechanism for the process. The standard methods give:

$$k_{\text{exp}}^{\infty} = \frac{\bar{k}T}{h}\frac{Q_{\ddagger}'}{Q_A}\,e^{-E_0^{\ddagger}/RT} \tag{5.12.1}$$

The value of $Q_{\ddagger}'$ depends on the type of complex assumed:

(i) If the activated complex is rigid and similar in structure to reactant A, then $Q_{\ddagger}' \sim Q_A$ and (5.12.1) becomes

$$k_{\text{exp}}^{\infty} \sim \frac{\bar{k}T}{h}\,e^{-E_0^{\ddagger}/RT}$$

This gives pre-exponential factors of 10^{12}–10^{13} s^{-1} as found experimentally for several cases.

(ii) The activated complex may be loose, as in the example:

$$C_2H_6 \rightarrow 2CH_3$$

where the complex is assumed to have almost free rotation about the C—C bond. In this case $Q_{\ddagger}' \gg Q_A$ and pre-exponential factors up to 10^{17} s^{-1} can be calculated.

This is as far as the basic ACT can be applied, but some of its postulates were incorporated into more detailed theories as described below.

5.13 RRKM Theory

In 1952 Marcus[21] developed an approach similar to the RRK treatment in considering free flow of energy between vibrational modes, but incorporating some ACT features. The method is known as RRKM and is based on the scheme:

$$A + A \underset{k_{-1}}{\overset{k_1}{\rightleftharpoons}} A^* + A$$

$$A^* \xrightarrow{k_2} A^{\ddagger} \longrightarrow \text{products}$$

k_1 was evaluated as a function of energy by a quantum statistical mechanical method and the elementary unimolecular step was formulated in ACT terms. A^* is an energized species as before, one which has sufficient energy for the unimolecular process but which must achieve the critical configuration of the activated complex, $A^{\ddagger}$, *en route* to reaction. The conceptual and mathematical difficulty now reaches a point where the student should turn to more advanced texts for a fuller treatment[22]. It is only

[21] R. A. Marcus, *J. Chem. Phys.*, **20**, 359 (1952).
[22] K. J. Laidler, *Theories of Chemical Reaction Rates*, McGraw-Hill, 1969.

appropriate here to try and convey something of the flavour of the RRKM method (and Slater's method in the next section) and to indicate its achievements.

The individual vibrational frequencies of A^* and $A^\ddagger$ are considered explicitly, the way normal mode vibrations contribute to reaction is taken into account and allowance is made for zero-point energies of all species. The starting point is again equation (5.11.4)

$$\mathrm{d}k_{\mathrm{exp}} = \frac{k_2(\mathrm{d}k_1/k_{-1})}{1 + k_2/k_{-1}[A]} \tag{5.11.4}$$

The RRKM treatment gives

$$\frac{\mathrm{d}k_1}{k_{-1}} = \frac{N^*(\varepsilon^*)\,\mathrm{e}^{-\varepsilon^*/\bar{k}T}}{q_\mathrm{a}}\,\mathrm{d}\varepsilon^* \tag{5.13.1}$$

$N^*(\varepsilon^*)$ is the energy density of the 'active' degrees of freedom—those which contribute to the breaking of bonds—as opposed to 'adiabatic' degrees of freedom which do not. ε^* is the 'active' energy in the activated complex, somewhat above the complex's zero-point energy, and q_a is the partition function corresponding to 'active' energy contribution.

k_2 is calculated by considering the relationship between the concentration of activated complexes and energized molecules:

$$k_2 = q_\mathrm{R}{}^\ddagger \frac{\int_0^\varepsilon N_2(\varepsilon_\mathrm{n}{}^\ddagger)}{q_\mathrm{R} N^*(\varepsilon^*) h}\,\mathrm{d}\varepsilon_\mathrm{n}{}^\ddagger \tag{5.13.2}$$

The active energy ε^* is the sum of $\varepsilon_\mathrm{t}{}^\ddagger$, for motion along the reaction coordinate, and $\varepsilon_\mathrm{n}{}^\ddagger$ which relates to vibrational modes and active rotations. $N_2(\varepsilon_\mathrm{n}{}^\ddagger)$ represents the energy density of quantum states at energy $\varepsilon_\mathrm{n}{}^\ddagger$ for vibrational modes. $q_\mathrm{R}{}^\ddagger/q_\mathrm{R}$ is the ratio of partition functions which allows for changes in moments of inertia in adiabatic rotations on forming the activated state from the energized molecule. Substitution of these terms into (5.11.5) and integration over the appropriate energy range gives the variation of k_{exp} with $[A]$, once again as a complicated function.

The RRKM method has been the most successful and commonly used theory of unimolecular reactions. Its application is illustrated in figure 5.13.

It can be shown that at high pressures the result reduces to

$$k_{\mathrm{exp}}^{\infty} = \frac{\bar{k}T}{h}\frac{Q_\ddagger{}'}{Q_A}\,\mathrm{e}^{-E_0{}^\ddagger/RT}$$

which is the same as the simple ACT result.

5.14 Slater's Theory

As opposed to statistical theories with energy flow between oscillators, Slater[23] postulated a *dynamic* theory without energy flow. Vibrations are assumed to be

[23] N. B. Slater, *Proc. Camb. Phil. Soc.*, **35**, 56 (1939).

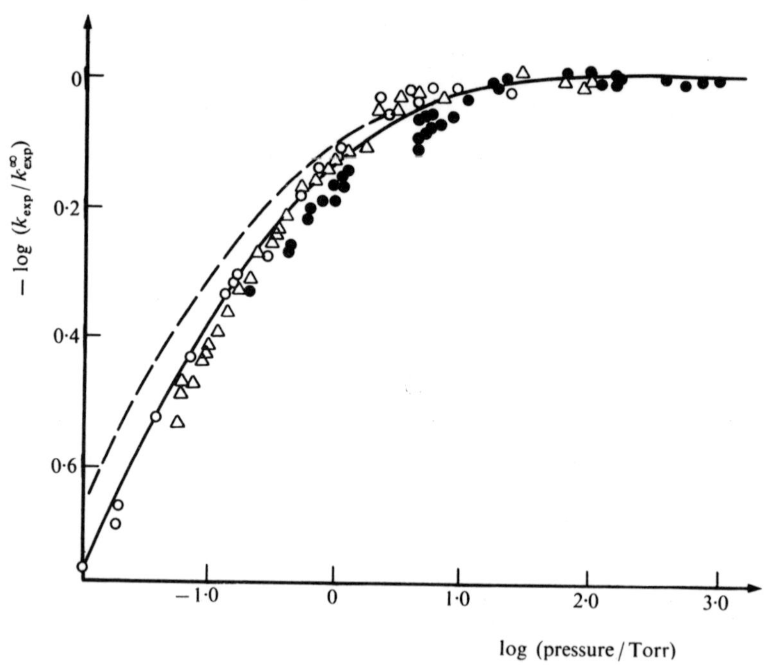

Figure 5.13
Comparison of experimental results and theory for decomposition of cyclobutane†. The full and broken lines are RRKM calculations based on slightly different values for k^{∞}_{exp}.

† P. J. Robinson and K. A. Holbrook, *Unimolecular Reactions*, Wiley Interscience, 1972, p. 248.

simple harmonic—which does not permit energy flow between modes—but when different modes come into phase the vibrational amplitude is changed. For suitable phase relationships the vibrational amplitude may extend beyond the critical length corresponding to reaction. This is illustrated in figure 5.14.

Slater gave a classical treatment in 1939 and later a quantum mechanical derivation. The mathematics are again complex but the variation of k_{exp} with $[A]$ can be calculated. The number of modes that may contribute to the critical reaction coordinate is *chosen* to give the best fit with experimental data, and the result is often approximately $\frac{1}{2}$ the maximum possible modes.

A simpler result emerges, again, for the high-pressure rate constant:

$$k^{\infty}_{exp} = \bar{\nu}\, e^{-\varepsilon^*/\bar{k}T}$$

ε^* is the minimum energy for the process as defined by the theory, and $\bar{\nu}$ is an average vibrational frequency—a weighted root-mean square of normal mode frequencies, where the weighting factors represent the contribution of each mode to displacing the critical coordinate.

Although the concept of reaction occurring when a critical coordinate is suitably extended is a realistic one, assuming no energy flow between modes is not. Various

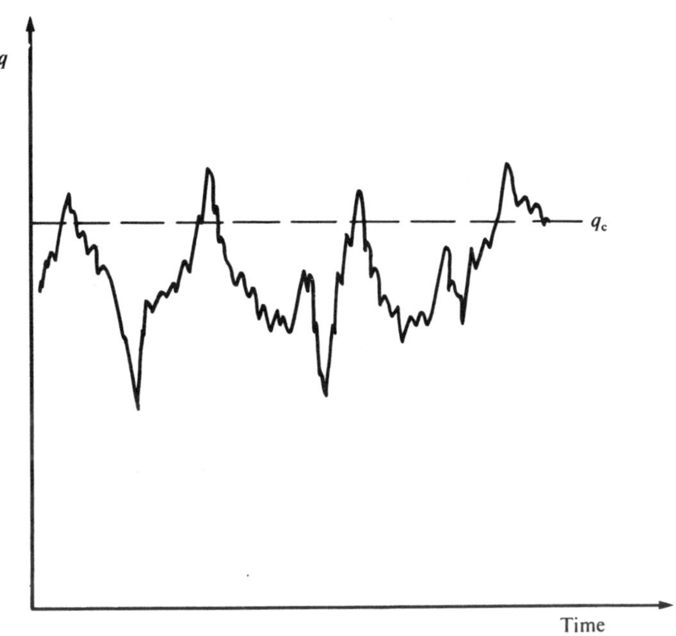

Figure 5.14
The variation of amplitude of an internal coordinate q with time. q_c is the critical value at which unimolecular reaction occurs.

attempts have been made to develop a theory along similar lines without the restriction on energy flow, by Gill and Laidler[24] for example.

Though considerable success has attended Slater's method, on the whole RRKM theory shows greater consistency with experimental results and is the theory most extensively used for predicting and correlating unimolecular rate parameters. RRKM methods are 'increasingly being used as a routine calculational technique rather than an esoteric theory for experts'[25].

Further Reading

E. A. Moelwyn-Hughes, *Physical Chemistry*, Pergamon, Oxford, 1957.

O. K. Rice, *Statistical Mechanics, Thermodynamics and Kinetics*, Freeman, London, 1967.

Comprehensive Chemical Kinetics, Vol. 2, (Eds C. H. Bamford and C. F. H. Tipper), Elsevier, 1969.

K. J. Laidler, *Theories of Chemical Reaction Rates*, McGraw-Hill, 1969.

[24] E. K. Gill and K. J. Laidler, *Proc. R. Soc.* (*London*), **A250**, 121 (1959); *Proc. R. Soc.* (*London*), **A251**, 66 (1959).

[25] P. J. Robinson, *Specialist Periodical Report: Reaction Kinetics*, Vol. 1, Chemical Society, London, 1975, p. 93.

J. C. Polanyi, *Acct. Chem. Res.*, **5**, 161 (1972).

D. L. Bunker, *Acct. Chem. Res.*, **7**, 195 (1974).

H. Eyring and E. M. Eyring, *Modern Chemical Kinetics*, Chapman and Hall, London, 1969.

H. S. Johnston, *Gas Phase Reaction Rate Theory*, Ronald Press, 1966.

H. S. Johnson and J. Buks, *Acct. Chem. Res.*, **5**, 327 (1972).

L. S. Kassel, *The Kinetics of Homogeneous Gas Reactions*, Chemical Catalog Co. Ltd., 1932.

C. N. Hinshelwood, *The Kinetics of Chemical Change*, Oxford University Press, London, 1940.

N. B. Slater, *Theory of Unimolecular Reactions*, Methuen, London, 1959.

P. J. Robinson and K. A. Holbrook, *Unimolecular Reactions*, Wiley Interscience, 1971.

Exercises

5.1

Calculate the number of collisions $cm^{-3}\ s^{-1}$ between A and B molecules for a mixture containing 100 Torr of each at 300 K. The molecular diameters of A and B are 0·3 and 0·4 nm respectively and the average relative velocity is $5 \times 10^2\ m\ s^{-1}$ at 300 K.

(Ans: $2{\cdot}3 \times 10^{27}$ collision $cm^{-3}\ s^{-1}$)

5.2

The rate constant for reaction between A and B, the same species as in exercise 5.1, is $1{\cdot}18 \times 10^5\ mol^{-1}\ cm^3\ s^{-1}$ at 300 K and the activation energy is 40 kJ mol^{-1}. Calculate the fraction of collisions at 300 K that occur with sufficient energy for reaction (on the SCT assumption) and evaluate the steric factor for the reaction.

(Ans: Fraction $= 10^{-7}$; steric factor $= 10^{-2}$)

5.3

Apply the BEBO method to plot the variation of potential energy with interatomic distance along the reaction path for the reaction $H_A + H_BH_C \rightarrow H_AH_B + H_C$. The lowest energy path corresponds to $n_{AB} + n_{BC} = 1$, and for the partial bonds: $E = 457{\cdot}4\ n^{1{\cdot}086}$ kJ mol^{-1}. The variation of bond order with interatomic distance, r (in nm), is given empirically by: $r = r_0 - 0{\cdot}026 \ln n$, where r_0 is the bond length in the molecule (0·074 nm). Choose 0·05 intervals for each n.

What is the height of the potential energy barrier?

(Ans: 26·5 kJ mol^{-1})

5.4

Write the ACT expression for the rate constant of a reaction between two non-linear species which form a non-linear activated complex, in terms of the partition functions for each degree of freedom of each species.

If the transmission coefficient is unity, $kT/h = 10^{12}\ s^{-1}$, all translational partition functions per degree of freedom are $10^8\ cm^{-1}$, all rotational partition functions are 10 and all vibrational partition functions are 1, show that the rate constant has the value $6 \times 10^8\ e^{-E_0^{\ddagger}/RT}\ mol^{-1}\ cm^3\ s^{-1}$.

5.5
A termolecular reaction between three diatomic species proceeds through a non-linear activated complex. If partition functions for each degree of freedom have the same value as in exercise 5.4 show that the rate constant has the value $3{\cdot}6 \times 10^8\ e^{-E_0^\ddagger/RT}$ $mol^{-2}\ cm^6\ s^{-1}$.

5.6
The reaction $2XY \rightarrow X_2 + Y_2$ proceeds via a non-linear activated complex $(XY)_2^\ddagger$. $E_0^\ddagger$ is zero, and the rate constant can be calculated from the ACT result with transmission coefficient unity if the expression is multiplied by 1/8 to allow for the effect of molecular symmetries. Expressions for the partition function for each degree of freedom are given in table 5.2. The functions can be evaluated from the following data:
$m_{XY} = 8{\cdot}47 \times 10^{-27}$ kg, $I_{XY} = 4{\cdot}3 \times 10^{-46}$ kg m^2. The product of the three moments of inertia of the complex is estimated as $(I_A I_B I_C) = 2{\cdot}20 \times 10^{-135}$ kg^3 m^6. The vibration frequency of XY is $2{\cdot}4 \times 10^{13}$ s^{-1} and the five vibrational frequencies of the complex are $(4{\cdot}5, 2{\cdot}4, 2{\cdot}1, 1{\cdot}8, 0{\cdot}6) \times 10^{13}$ s^{-1} respectively.

Show that at 400 K the rate constant is $1{\cdot}93 \times 10^{10}$ mol^{-1} cm^3 s^{-1}.

5.7
The values of the experimental rate constant for a unimolecular reaction at a series of temperatures are shown below

T/K	500	515	535	545	600
$k/10^{-4}$ s^{-1}	0·23	0·71	2·9	5·5	140

Calculate the entropy and enthalpy of activation for the average temperature in this range.

(Ans: $\Delta S_0^\ddagger = -27$ J mol^{-1} K^{-1}; $\Delta H_0^\ddagger = 155$ kJ mol^{-1})

5.8
Show that for a termolecular atom combination, the energy transfer mechanism given in section 5.8 is consistent with third-order kinetics.

5.9
At 800 K the high-pressure limiting value for the rate constant of a unimolecular reaction is 5×10^{-4} s^{-1}. At the same temperature the first-order rate constant falls to half this value at a pressure of 2×10^{-2} Torr. Calculate the rate constant for the molecular activation step.

(Ans: $1{\cdot}14 \times 10^6$ mol^{-1} cm^3 s^{-1})

5.10
For the unimolecular reaction of exercise 5.9 the molecular activation step has an 'activation energy' of 240 kJ mol^{-1}. Confirm that the pre-exponential factor is many orders of magnitude greater than allowed by the SCT ($Z_{AA} = 1 \times 10^{14}$ mol^{-1} cm^3 s^{-1} at 800 K) and calculate the approximate number of vibrational modes that are effective in the energy exchange, according to Hinshelwood's model.

(Ans: $n \sim 9$)

Chapter 6: Radical Reactions, Non-Chain and Straight-Chain Reactions

Over the first decades of this century it became apparent that most gas phase reactions proceeded by a series of elementary radical reactions, but for many years it was thought that certain reactions were part molecular, and others entirely molecular in nature. Cumulative experience with both conventional kinetic methods, discussed in chapter 3, and the techniques for fast reactions described in chapter 4, has now established that virtually all gas phase reactions have radical mechanisms (at temperatures below the onset of ionic processes). The number of elementary radical reactions that has been identified and investigated is truly immense, but fortunately they fall into relatively few categories, which will be presented here, accompanied by a few general observations on each type. Even for reactions where rate constants and activation energies have not been measured it is possible to speculate on their rates. A wide variety of bond dissociation energies is known from thermochemical and spectroscopic data, allowing the calculation of heats of reaction. (Some useful values are given in table 6.1[1].) There is no theoretical relationship between heat of reaction and activation energy but, as pointed out in section 5.4, there are many such empirical relationships. They cover the range from highly exothermic reactions, with almost zero activation energy, to endothermic reactions with activation energies which are greater than their endothermicities.

This account of elementary reaction types will be followed by a description of some non-chain gas reactions and the experimental and mechanistic features of chain reactions. The chapter concludes with a detailed discussion of some important unbranched-chain reactions.

[1] These values are derived from a variety of sources, but mainly V. I. Vedeneyev *et al.*, *Bond Energies, Ionisation Potentials and Electron Affinities*, Arnold, London, 1966, and J. A. Kerr, *Chem. Rev.*, **66**, 465 (1966). Most should be accurate to within ± 8 kJ mol^{-1}.

Table 6.1
Bond dissociation energies D (at 298 K).

Bond	$D/\mathrm{kJ\ mol^{-1}}$	Bond	$D/\mathrm{kJ\ mol^{-1}}$
$HC{\equiv}CH$	963	F—F	165
$H_2C{=}CH_2$	698	Cl—Cl	242
$H_3C—CH_3$	368	Br—Br	193
$H_3C—CH_2CH_3$	356	I—I	151
$H_3C—C^{\cdot}O$	46		
$H_3C—COCH_3$	338	H—F	566
		H—Cl	437
$H—CH_3$	435	H—Br	370
$H—CH_2CH_3$	410	H—I	298
$H—CH_2CH_2CH_3$	416		
$H—CH_2C^{\cdot}H_2$	163	$F—CF_3$	506
		$Cl—CCl_3$	305
$H—CH(CH_3)_2$	395	$Br—CBr_3$	208
$H—C(CH_3)_3$	380		
$H—CHCH_2$	440	$F—CH_3$	451
H—CCH	481	$Cl—CH_3$	350
$H—C^{\cdot}O$	75	$Br—CH_3$	292
H—CHO	368	$I—CH_3$	235
$H—COCH_3$	368		
		$N{\equiv}N$	945
H—H	435	N=O	640
H—D	439	ON—O	306
D—D	443	$H—NH_2$	440
		$H_2N—NH_2$	250
$O^{\cdot}—H$	428	$H_3C—NH_2$	334
$H—O_2^{\cdot}$	201		
H—OH	498	C=O	1074
HO—OH	214	OC=O	532
RO—OH	~165	$H_3\dot{C}—OH$	380
$CH_3O—OCH_3$	150		
		O=O	498
		$O_2—O$	105

6.1 Classification of Elementary Radical Reactions

(I) molecular fission

$$AB \rightarrow A^{\cdot} + B^{\cdot}$$

Unimolecular reactions of this type have been discussed in detail in sections 5.9–14, and some examples, with rate parameters, are given in table 5.5. In such reactions a bond in a stable molecule is broken so they are highly endothermic and of high activation energy, usually in the range 160–380 kJ mol^{-1}. Thus, although pre-exponential factors are often high, about 10^{16} s^{-1}, such steps are slow and the complex reactions in which they provide the source of radicals are slow unless the temperature is very high. If the parent molecule possesses a suitable absorption spectrum this slow step for the initial formation of radicals may be replaced by the corresponding photochemical process:

$$AB + h\nu \rightarrow A^{\cdot} + B^{\cdot}$$

which has zero activation energy in the conventional sense. Such photochemical processes are surveyed in detail in chapter 2.

(II) radical fission

$$ABC^{\cdot} \rightarrow AB + C^{\cdot}$$

This is the radical analogue of type I above and again some examples are given in table 5.5.

Pre-exponential factors are generally in the range 10^{13}–10^{15} s^{-1}, and activation energies are high. In fact activation energies are somewhat lower than with type I, since structural re-arrangement in molecular fragment AB gives stronger bonding, and this compensates for some of the energy required to break the initial bond, e.g.

$$C_2H_5^{\cdot} \rightarrow C_2H_4 + H^{\cdot}$$

(III) isomerization

$$A_1^{\cdot} \rightarrow A_2^{\cdot}$$

This unimolecular process is possible with large radicals that can undergo structural or geometrical re-arrangement. Some examples are:

$$\text{cyclo-}(H_2C\text{—}CH_2\text{—}\dot{C}H) \rightarrow \dot{C}H_2 - CH = CH_2$$

$$CH_3CH_2CH_2CH_2C^{\cdot}H_2 \rightarrow CH_3C^{\cdot}HCH_2CH_2CH_3$$

(IV) radical–molecule addition

$$AB + C^{\cdot} \rightarrow ABC^{\cdot}$$

The most common examples of this bimolecular reaction are additions to double bonds, as illustrated in table 6.2. The result is a higher molecular weight radical which may undergo a similar reaction to give a yet heavier species—processes of this type are important in polymerizations. The reactions are exothermic and of low activation energy, about 4 kJ mol^{-1} for atomic and 30 kJ mol^{-1} for polyatomic radical additions. Pre-exponential factors are usually of the order of 10^{11} mol^{-1} cm^3 s^{-1}.

(V) exchange or transfer reactions

$$A^{\cdot} + BC \rightarrow AB + C^{\cdot}$$

This bimolecular reaction is the most common general category. $A^{\cdot}$ may be an atom or polyatomic radical and it abstracts an atom B (very frequently a hydrogen atom) leaving radical $C^{\cdot}$. Only rarely is the exchanged species B a polyatomic radical. A selection of examples is given in table 6.2. The heat of reaction depends on whether the newly formed bond is of higher or lower dissociation energy than the bond broken. Examples vary from highly exothermic reactions with almost zero activation energies to endothermic reactions of very high activation energy. Typical pre-exponential factors are 10^{13}–10^{14} mol^{-1} cm^3 s^{-1} for abstraction by atoms, and 10^{11}–10^{12} mol^{-1} cm^3 s^{-1} for radicals.

Table 6.2
Experimental rate parameters for bimolecular and termolecular radical reactions.

Reaction	$\log_{10}(A_{exp}/mol^{-1}\ cm^3\ s^{-1})$	$E_{exp}/kJ\ mol^{-1}$
Radical–molecule addition		
$H^{\cdot} + C_2H_4 \rightarrow C_2H_5^{\cdot}$	11·0	~0
$CH_3^{\cdot} + C_2H_4 \rightarrow C_3H_7^{\cdot}$	11·0	29
$C_2H_5^{\cdot} + C_2H_4 \rightarrow C_4H_9^{\cdot}$	11·0	29
Exchange reactions		
$CH_3^{\cdot} + H_2 \rightarrow CH_4 + H^{\cdot}$	11·7	44
$CH_3^{\cdot} + C_2H_6 \rightarrow CH_4 + C_2H_5^{\cdot}$	11·2	44
$CH_3^{\cdot} + C_3H_8 \rightarrow CH_4 + C_3H_7^{\cdot}$	11·4	41
$CH_3^{\cdot} + (CH_3)_3CH \rightarrow CH_4 + (CH_3)_3C^{\cdot}$	10·9	32
$CH_3^{\cdot} + CH_3CHO \rightarrow CH_4 + CH_3CO^{\cdot}$	12·0	33
$CH_3^{\cdot} + (CH_3)_2CO \rightarrow CH_4 + CH_3COCH_2^{\cdot}$	11·4	40
$C_2H_5^{\cdot} + H_2 \rightarrow C_2H_6 + H^{\cdot}$	11·8	47
$Cl^{\cdot} + H_2 \rightarrow HCl + H^{\cdot}$	13·9	23
$Br^{\cdot} + H_2 \rightarrow HBr + H^{\cdot}$	13·9	72
$I^{\cdot} + H_2 \rightarrow HI + H^{\cdot}$	14·1	140
$H^{\cdot} + HCl \rightarrow H_2 + Cl^{\cdot}$	13·6	19
$H^{\cdot} + HBr \rightarrow H_2 + Br^{\cdot}$	13·2	4
$H^{\cdot} + HI \rightarrow H_2 + I^{\cdot}$	12·0	~0
$H^{\cdot} + H_2 \rightarrow H_2 + H^{\cdot}$	13·7	31
$H^{\cdot} + C_2H_6 \rightarrow H_2 + C_2H_5^{\cdot}$	14·1	41
$H^{\cdot} + C_3H_8 \rightarrow H_2 + C_3H_7^{\cdot}$	13·9	35
$H^{\cdot} + NO_2 \rightarrow NO + OH^{\cdot}$	13·5	~0
$H^{\cdot} + O_2 \rightarrow OH^{\cdot} + O^{\cdot}$	14·5	71
$O^{\cdot\cdot} + H_2 \rightarrow OH^{\cdot} + H^{\cdot}$	13·5	40
$O^{\cdot\cdot} + NO_2 \rightarrow NO + O_2$	13·3	4·6
$O_3 + NO \rightarrow NO_2 + O_2$[a]	12·0	10
Radical combination with disproportionation		
$2C_2H_5^{\cdot} \rightarrow C_2H_4 + C_2H_6$	12·4	~0
Radical combination		
$2CH_3^{\cdot} \rightarrow C_2H_6$	13·3	~0
$2C_2H_5^{\cdot} \rightarrow C_4H_{10}$	13·2	~0
$2CCl_3^{\cdot} \rightarrow C_2Cl_6$	13·5	~0
$Cl^{\cdot} + COCl^{\cdot} \rightarrow COCl_2$	14·6	3·3
	$\log_{10}(A/mol^{-2}\ cm^6\ s^{-1})$	
$H^{\cdot} + H^{\cdot} + H_2 \rightarrow H_2 + H_2$	16·0	~0
$O^{\cdot\cdot} + O^{\cdot\cdot} + O_2 \rightarrow O_2 + O_2$	15·0	~0
$O^{\cdot\cdot} + O_2 + Ar \rightarrow O_3 + Ar$	12·5	−9·6
$H^{\cdot} + O_2 + Ar \rightarrow HO_2 + Ar$	14·7	−6·7
$I^{\cdot} + I^{\cdot} + Ar \rightarrow I_2 + Ar$[b]	15·8	−5·4

[a] Not strictly a radical reaction.
[b] For other third bodies in iodine recombination see table 5.4.

(VI) radical combination with disproportionation

This occurs when the bimolecular reaction of complex radicals can result in two stable molecules, e.g.:

$$C_2H_5^{\cdot} + C_2H_5^{\cdot} \rightarrow C_2H_6 + C_2H_4$$

(VII) radical combination

$$A^{\cdot} + B^{\cdot} \rightarrow AB$$

As the new bond is formed, its dissociation energy is released. Thus the new species will dissociate again unless some energy is transferred intramolecularly to other vibrational modes. This reaction also competes with the possibility of disproportionation as above. The formation of a new bond means very high exothermicity and virtually zero activation energy. Pre-exponential factors are about 10^{13}–10^{14} mol^{-1} cm^3 s^{-1}.

For the combination of very small radicals or atoms a third body is required to remove some of the exothermicity. In such cases the reaction is third order and best represented:

$$A^{\cdot} + B^{\cdot} + M \rightarrow AB + M$$

Examples of termolecular processes are given in tables 5.4 and 6.2 which illustrate some features discussed at length in section 5.8, particularly the negative activation energies and varying third body efficiencies.

6.2 Non-Chain Reactions

Gas reactions can be extremely complicated but some relatively simple cases are known. For example radical reactions of types I, II and VII occur in the *decomposition of di-t-butyl peroxide*, a useful source of methyl radicals.

$$(CH_3)_3CO - OC(CH_3)_3 \rightarrow 2(CH_3)_3CO^{\cdot}$$

$$(CH_3)_3CO^{\cdot} \rightarrow (CH_3)_2CO + CH_3^{\cdot}$$

$$2CH_3^{\cdot} \rightarrow C_2H_6$$

$$\text{overall:} \quad (CH_3)_3COOC(CH_3)_3 \rightarrow 2(CH_3)_2CO + C_2H_6$$

A reaction which perplexed kineticists for many years is the much-studied ‘unimolecular’ *decomposition of nitrogen pentoxide.*

The overall reaction is:

$$N_2O_5 \rightarrow 2NO_2 + \tfrac{1}{2}O_2$$

and the overall rate is $-d[N_2O_5]/dt = k[N_2O_5]$.

Ogg[2] proposed the mechanism, now widely accepted:

$$N_2O_5 \xrightarrow{k_1} NO_2 + NO_3^{\cdot}$$

$$NO_2 + NO_3^{\cdot} \xrightarrow{k_2} N_2O_5$$

$$NO_2 + NO_3^{\cdot} \xrightarrow{k_3} NO + O_2 + NO_2$$

$$NO + NO_3^{\cdot} \xrightarrow{k_4} 2NO_2$$

[2] R. A. Ogg, *J. Chem. Phys.*, **15**, 337 (1947); *J. Chem. Phys.*, **18**, 572 (1950).

Applying the stationary-state hypothesis to $[NO_3]$ and $[NO]$ gives

$$-\frac{d[N_2O_5]}{dt} = \frac{2k_3k_1}{k_2 + 2k_3}[N_2O_5]$$

So the first-order rate constant, the source of much speculation, is a composite of k_1, k_2 and k_3.

The *hydrogen–iodine* and reverse reactions were the subject of the first comprehensive kinetic investigation of a gas reaction. Bodenstein[3] showed that the slow reaction:

$$H_2 + I_2 \rightarrow 2HI$$

was second order. For many years this was held to be a prime example of a simple bimolecular reaction, with a clearly defined four-centre activated complex. Later it was found that above 800 K the reaction proceeded by a predominantly chain mechanism, and in 1967 Sullivan[4] showed that the low-temperature non-chain reaction also had a radical mechanism.

Iodine atoms are formed:

$$I_2 + M \xrightarrow{k_1} 2I^{\cdot} + M$$

The subsequent iodine atom reaction:

$$I^{\cdot} + H_2 \rightarrow HI + H^{\cdot}$$

is highly endothermic, and only at temperatures approaching 800 K is it fast enough to compete with:

$$2I^{\cdot} + M \xrightarrow{k_2} I_2 + M$$

The slow formation of HI is due to a similar process where the third body is H_2:

$$2I^{\cdot} + H_2 \xrightarrow{k_3} 2HI \qquad (6.2.I)$$

This termolecular step probably proceeds through an $I..H_2$ intermediate (see section 5.8). With this mechanism, taking $k_2 \gg k_3$, it is easily shown that HI formation is second order:

$$\frac{d[HI]}{dt} = \frac{2k_3k_1}{k_2}[H_2][I_2]$$

A subject of great concern at present is *the ozone system*, as the reactions by which ozone is formed and removed in the lower region of the stratosphere are crucially important to mankind's survival. The stratospheric ozone absorbs ultraviolet radiation from the sun, radiation that could otherwise have a severe effect on life at the surface:

$$O_3 + h\nu \rightarrow O_2 + O^{\cdot\cdot} \qquad (6.2.II)$$

In the stratosphere, atmospheric pressure is very low—the concentration of oxygen molecules has fallen to about 10^{-7} mol cm^{-3} at 30 km altitude. Oxygen atoms are

[3] M. Bodenstein, *Z. Phys. Chem.*, **13**, 56 (1894); *Z. Phys. Chem.*, **22**, 1 (1897); *Z. Phys. Chem.*, **29**, 295 (1899).

[4] J. H. Sullivan, *J. Chem. Phys.*, **46**, 73 (1967).

formed mainly by photolysis of oxygen at higher altitudes, and their concentration is relatively high. The more important reactions in the oxygen–ozone system are, in addition to (6.2.II), those given below. The situation is, in fact, somewhat complicated by the participation of excited states of these species[5].

$$O_2 + h\nu \rightarrow O^{\cdot\cdot} + O^{\cdot\cdot} \tag{6.2.III}$$

$$O^{\cdot\cdot} + O_2 + M \rightarrow O_3 + M \tag{6.2.IV}$$

$$O^{\cdot\cdot} + O + M \rightarrow O_2 + M \tag{6.2.V}$$

$$O^{\cdot\cdot} + O_3 \rightarrow O_2 + O_2 \tag{6.2.VI}$$

An approximate photochemical equilibrium exists for ozone, but for some years severe apprehension has been expressed about the possible effect on the ozone concentration of nitric oxide emitted by the engines of supersonic planes, particularly passenger transports such as Concorde which would fly in the stratosphere. Already calculations based on the above reaction scheme yield ozone concentrations higher than those observed, and such depletion would be enhanced by the reaction

$$NO + O_3 \rightarrow NO_2 + O_2$$

Calculations including this reaction in the scheme suggest that the effect of nitric oxide may be catastrophic[6]. The evaluation in the laboratory of rate constants for these reactions is clearly of far more than academic interest, and in chapter 4 we have seen examples of suitable techniques by which this has been accomplished for reactions (6.2.IV) and (6.2.V).

Further alarm has been raised very recently over the accumulation in the stratosphere of some freons, very stable fluoro-chloro-methanes, which are essential ingredients in the ever-widening range of refrigerants and commercial aerosol products. These have now reached concentrations where they may contribute significantly to ozone depletion[7], photolysis in the stratosphere, e.g.

$$CF_2Cl_2 + h\nu \rightarrow CF_2Cl^{\cdot} + Cl^{\cdot}$$

being followed by:

$$Cl^{\cdot} + O_3 \rightarrow ClO^{\cdot} + O_2$$

6.3 Chain Reactions: Experimental Characteristics

Radicals often participate in two or more consecutive exchange reactions, for example:

$$A^{\cdot} + L \rightarrow P + B^{\cdot}$$

$$B^{\cdot} + M \rightarrow Q + A^{\cdot}$$

where $A^{\cdot}$ and $B^{\cdot}$ are radicals, L and M reactant molecules, and P and Q products.

This introduces the important possibility that the initial radical ($A^{\cdot}$) is regenerated and the series of reactions can be repeated either until all reactants are consumed or, more

[5] R. P. Wayne, *Photochemistry*, Butterworths, London, 1970.
[6] H. Johnston, *Science*, **173**, 517 (1971).
[7] M. J. Molena and F. S. Rowland, *Nature*, **249**, 810 (1974).

realistically, until $A^{\cdot}$ or $B^{\cdot}$ undergoes a competing reaction in which radicals are removed. Hence the formation of radical $A^{\cdot}$ initiates a chain of reactions in which many product molecules are formed. *Chain reactions* exhibit some characteristic experimental features[8]:

Steady rate

The rate of a non-chain reaction is highest at the start of reaction and falls with time. The rate of a chain reaction is zero initially and builds up to a steady value which ultimately falls as reactants are consumed. Such a steady rate is illustrated in figure 7.2.

Inhibition and sensitization

The reaction rate depends on the radical concentrations and small concentrations of additives which alter the radical concentrations can have a marked effect on the rate. A classic example is the addition of nitric oxide to hydrocarbon systems where the nitric oxide combines directly with many radicals and so *inhibits* the reaction. Though inhibition is a useful diagnostic test for radical chains, great care must be exercised in any detailed mechanistic deductions, as many kineticists know to their cost in the case of nitric oxide. Increasing the reaction rate with an additive is called *sensitization*. For example, the pyrolysis of acetaldehydes involves the participation of acetyl radicals, and addition of a small amount of biacetyl sensitizes the reaction through the extra acetyl radicals it provides.

High quantum yield

Where complex reactions are initiated photochemically it is usually simple to measure the quantum yield (see section 2.6). In non-chain systems the quantum yield rarely exceeds 2, and high values indicate chain reactions. A good example is the hydrogen–chlorine photochemical reaction whose quantum yield is of the order of 10^6.

Effect of reaction vessel surface

Many chain reactions depend critically on the surface of the reaction vessel. This can be demonstrated by the effect of changing the surface: volume ratio on reaction rate. Such studies can be carried out conveniently by using cylindrical or spherical vessels of different diameters or, more drastically, by packing a vessel with rods or spheres of appropriate material. The surface reactions which participate are surface catalysed molecular dissociations or radical combinations. These reactions are also influenced by applying different coatings to the surface—lead oxide and potassium chloride, for example, having widely different efficiencies for surface radical combinations.

Effect of inert gases

The rates of chain reactions are influenced by the addition of inert gases, not only the rare gases but N_2, CO_2 and others that undergo no chemical change in the system. These gases may act as third bodies in termolecular radical combination reactions or may reduce the rate of radical diffusion to the vessel surface simply by increasing the total pressure.

[8] These are described in more detail in F. S. Dainton, *Chain Reactions*, 2nd ed., Methuen, London, 1966.

Explosion

At a critical conjunction of temperature and reactant pressure some chain reactions undergo an exponential increase in reaction rate to an unmeasurably high value, as illustrated in figure 7.2. This is a kinetic definition of explosion, and it can be accomplished under conditions not too violent for the survival of the apparatus! The influence of pressure and temperature is such that very small changes in either can result in a dramatic change from non-explosive to explosive rates, and well-defined *explosion limits* may be drawn for these critical temperatures and pressures. An example is given in figure 7.3.

Explosion can also occur by a largely thermal route. Fast exothermic reactions can give a significant rise in temperature which stimulates a related accelerating increase in reaction rate leading to explosion. This is discussed in detail in section 7.1.

6.4 Elementary Steps in Chain Reactions

All the diverse experimental features outlined above are explicable in terms of chain reaction mechanisms. Although all chain reactions differ, there are several crucial common factors exhibited by the series of radical reactions which comprise their mechanisms, and these general reaction types are considered below. Subsequently several chain reactions are described in detail, highlighting the origin of the salient experimental characteristics and illustrating the way detailed kinetic information is deduced from the results. The study of chain reactions is a field of sufficient importance for the growth of its own terminology. Radicals involved in the chain are referred to as *chain carriers* and the radical reactions are accorded a somewhat different classification presented below[9]. The equivalent reaction types, as defined in section 6.1, will be quoted where appropriate.

(i) Initiation steps

In these, radicals are formed from stable molecules (type I)

$$C_2H_6 \rightarrow 2CH_3^{\cdot}$$

$$Br_2 + M \rightarrow 2Br^{\cdot} + M$$

These are highly endothermic, high activation energy reactions and photochemical initiation is a useful alternative:

$$Br_2 + h\nu \rightarrow 2Br^{\cdot}$$

The quoted examples are for homogeneous initiation. Heterogeneous initiation is also possible, and it predominates where the step of lowest activation energy is a surface catalysed decomposition. This may occur, for example, in the chain reaction between hydrogen and oxygen, where a surface reaction giving hydroxyl radicals probably provides initiation.

(ii) Propagation steps

These represent the essential feature that separates chain from non-chain reactions. A product is formed and another radical chain carrier is generated to react again and so

[9] For a more detailed exposition, see F. S. Dainton, *Chain Reactions*, 2nd ed., Methuen, London, 1966.

propagate the chain. In such reactions there is no net change in the number of radicals in the system, e.g.

$C_2H_5^{\cdot} \rightarrow C_2H_4 + H^{\cdot}$ (type II)

$CH_3^{\cdot} + C_2H_4 \rightarrow C_3H_7^{\cdot}$ (type IV)

$Cl^{\cdot} + H_2 \rightarrow HCl + H^{\cdot}$ (type V) (6.4.I)

$Br^{\cdot} + H_2 \rightarrow HBr + H^{\cdot}$ (6.4.II)

The rate of propagation has a decisive effect on the overall reaction rate. The hydrogen–chlorine reaction is much faster than the hydrogen–bromine reaction and one important contributing factor is the relative rate of the corresponding propagation steps (6.4.I) and (6.4.II) above. That for the chlorine reactions, (6.4.I), is only 5 kJ mol^{-1} endothermic, with an activation energy of 23 kJ mol^{-1}, whereas that for bromine is 69 kJ mol^{-1} endothermic with an activation energy of 77 kJ mol^{-1}, and (6.4.I) is much faster than (6.4.II).

(iii) Termination steps

Reactions in which radicals are removed and chains are broken may be homogeneous or heterogeneous:

(a) Homogeneous, e.g.

$2C_2H_5^{\cdot} \rightarrow C_2H_4 + C_2H_6$ (type VI)

$2C_2H_5^{\cdot} \rightarrow C_4H_{10}$ (type VII)

$2Br^{\cdot} + M \rightarrow Br_2 + M$ (type VII)

Termination steps compete with propagation steps for radicals. The homogeneous termination reactions have virtually zero activation energy, whereas propagation reactions have significant activation energies and propagation increases in relative importance as temperature increases. The termination steps are kinetically second order in radical concentration whereas propagation reactions are first order in this respect, so their relative rates vary with radical concentration.

(b) Heterogeneous termination

With efficient surface combination of radicals the rate determining part of the termination process is diffusion of the radicals to the vessel wall, e.g.:

$$H^{\cdot} \xrightarrow{k_D} \text{wall} \ (\rightarrow \tfrac{1}{2}H_2)$$

In such cases heterogeneous termination is kinetically first order. The 'rate constant' for diffusion, k_D, is unlike a reaction rate constant. It may be expressed, for a cylindrical or spherical vessel diameter d:

$$k_D = \frac{\tau D_0}{d^2 P}$$

τ is a numerical constant, D_0 the diffusion coefficient for unit pressure, and P the pressure. This rate constant is inversely proportional to pressure, so at very low pressure heterogeneous termination may dominate whereas at high pressure a competing homogeneous termination step may be more important. The rate of heterogeneous termination depends on vessel diameter so its influence is less in larger vessels. In addition the diffusion coefficient has a $T^{1/2}$ temperature dependence, and competing propagation reactions with activation energies are favoured as temperature increases.

The relative rate of competing propagation and termination steps is a critical influence on the overall reaction. A reflection of this is the *chain length*:

$$\text{Chain length} = \frac{\text{number of reactant molecules removed (or products formed)}}{\text{number of radicals formed in initiation step}}$$

The faster the propagation steps relative to termination, the greater will be the chain length. In photochemical systems the number of radicals formed in initiation steps can be deduced from the number of quanta absorbed, and there is a direct relation between chain length and quantum yield.

(iv) Branching steps

These are reactions in which the number of radicals increases, e.g.

$$H^{\cdot} + O_2 \rightarrow OH^{\cdot} + O^{\cdot\cdot} \qquad (6.4.III)$$

Chain reactions with such steps are referred to as *branched-chain reactions* as opposed to *straight-chain reactions*. Branching can lead to a continuing multiplication in the numbers of radicals present, an exponential rise in reaction rate, and explosion. Whether this occurs is determined by competition between branching and termination. Branching reactions are endothermic—in the example above the bond broken is stronger than the new bond formed and the reaction is 71 kJ mol^{-1} endothermic, with a similar activation energy. Termination steps show comparatively little variation with temperature, so although branching is slower at low temperature it may dominate at high temperature. For example, with the hydrogen–oxygen reaction, where (6.4.III) is the branching reaction, explosion only occurs above about 700 K.

Several types of branching reaction may be categorized. That quoted above is an example of a *normal* branching reaction. In some cases branching can result from the dissociation of a product molecule with a weak bond. This 'delayed' branching is called *degenerate*, e.g. in the oxidation of hydrocarbons hydroperoxy products can decompose:

$$R\text{OOH} \rightarrow R\text{O}^{\cdot} + \text{OH}^{\cdot}$$

In the combustion of carbon monoxide the chain carrier is an excited electronic state of carbon dioxide which causes branching by transferring energy to a molecule which dissociates:

$$CO_2^* + O_2 \rightarrow CO_2 + O^{\cdot\cdot} + O^{\cdot\cdot}$$

This is called *energy* branching.

Two further terms may be introduced here. Reactions which are first order in radicals, especially branching or termination steps, are called *linear*. If second order in radicals they are termed *quadratic*.

6.5 Unbranched-Chain Reactions: Hydrogen–Bromine Reaction

The pyrolysis of hydrocarbons (section 6.8) is a prominent example of an industrially important reaction. Hydrogen–halogen reactions may appear to be of comparatively academic interest but they played a central role in the history of kinetics. It was in these systems that gaseous radical reactions were first identified and some essential features of chain processes established. It was through understanding these relatively straightforward cases that interpretation of the more complicated systems became possible.

In 1906, Bodenstein[10] determined the experimental rate equation for the thermal reaction

$$H_2 + Br_2 \rightarrow 2HBr$$

finding the first example of a complex rate equation in gas kinetics:

$$\frac{d[HBr]}{dt} = \frac{k[Br_2]^{1/2}[H_2]}{1 + k_i[HBr]/[Br_2]} \qquad (6.5.1)$$

The reaction is inhibited by the accumulating concentration of product HBr described quantitatively by the value of 'inhibition constant', k_i in (6.5.1). The variation of rate with temperature showed that the activation energies for the 'overall' rate constant, k, and inhibition constant, k_i, were of the order of 170 kJ mol^{-1} and about 0 respectively.

The reaction mechanism was not established until 1919 when Christiansen, Herzfeld and Polanyi[11], independently, proposed a radical chain. To maintain a historical perspective it should be remembered that there was, at that time, no direct evidence that gaseous free radicals existed and virtually nothing was known of their reactions. The discussion here will, however, employ some kinetic hindsight in presenting a reasonable mechanism. The mechanism must be consistent with the experimental rate equation and it will then be possible to illustrate the way rate constants and activation energies for elementary reactions are deduced from the kinetic data.

Mechanism

For the thermal reactions at typical temperatures of around 500 K and pressures about 200 Torr, $\sim 2{\cdot}7 \times 10^4$ N m^{-2}, initiation and termination should be homogeneous. The Br—Br bond is much weaker than the H—H and its breaking will provide initiation. The chain reaction then proceeds:

Initiation	$Br_2 + M \xrightarrow{k_1} 2Br^{\cdot} + M$	(6.5.I)
Propagation	$Br^{\cdot} + H_2 \xrightarrow{k_2} HBr + H^{\cdot}$	(6.5.II)
	$H^{\cdot} + Br_2 \xrightarrow{k_3} HBr + Br^{\cdot}$	(6.5.III)
Termination	$2Br^{\cdot} + M \xrightarrow{k_4} Br_2 + M$	(6.5.IV)

[10] M. Bodenstein and S. C. Lind, *Z. Phys. Chem.*, **57**, 168 (1907).
[11] J. A. Christiansen, *K. Danske Vidensk. Selsk.*, **i**, No. 14 (1919); K. F. Herzfeld, *Z. Elektrochem.*, **25**, 301 (1919); M. Polanyi, *Z. Elektrochem.*, **26**, 50 (1920).

Inspection of the propagation steps shows that whereas (6.5.III) is 173 kJ mol^{-1} exothermic and very fast, (6.5.II) is 69 kJ mol^{-1} endothermic, with high activation energy, and so relatively slow. The reverse of (6.5.II):

$$\text{Inhibition} \qquad H^{\cdot} + HBr \xrightarrow{k_5} H_2 + Br^{\cdot} \qquad (6.5.V)$$

is correspondingly exothermic and fast. It provides inhibition by competing with propagation step (6.5.III) for H atoms once HBr concentration builds up.

Rate equation

From the mechanism, the rate of product formation is:

$$d[HBr]/dt = k_2[Br][H_2] + k_3[H][Br_2] - k_5[H][HBr] \qquad (6.5.2)$$

The experimental rate equation is expressed in terms of measurable concentrations, $[H_2]$, $[Br_2]$ and [HBr] so radical concentrations [H], [Br], must be eliminated from (6.5.2). Assuming that the steady state is set up rapidly—this can be confirmed when rate constants are known—the steady-state hypothesis can be applied to H and Br concentrations:

$$d[Br]/dt = 2k_1[Br_2][M] - k_2[Br][H_2] + k_3[H][Br_2] - 2k_4[Br]^2[M] + k_5[H][HBr] = 0 \qquad (6.5.3)$$

$$d[H]/dt = k_2[Br][H_2] - k_3[H][Br_2] - k_5[H][HBr] = 0 \qquad (6.5.4)$$

There are two simultaneous equations for two 'unknowns', [H], [Br] and from (6.5.4) and (6.5.3):

$$k_1[Br_2][M] = k_4[Br]^2[M] \qquad (6.5.5)$$

This is equivalent to *rate of initiation* = *rate of termination*, a generally valid result for unbranched-chain reactions in the steady state. (6.5.5) may be rewritten:

$$[Br]_{ss} = \{(k_1/k_4)[Br_2]\}^{1/2} = K_{Br}{}^{1/2}[Br_2]^{1/2} \qquad (6.5.6)$$

where K_{Br} is the bromine dissociation constant. So in the thermal system the steady-state concentration of Br is its equilibrium concentration. Substitution of (6.5.6) into equation (6.5.4) gives:

$$[H]_{ss} = \frac{k_2 K_{Br}{}^{1/2}[Br_2]^{1/2}[H_2]}{k_3[Br_2] + k_5[HBr]} \qquad (6.5.7)$$

From (6.5.2) and (6.5.4), the rate equation may be written

$$d[HBr]/dt = 2k_3[H][Br_2]$$

which from (6.5.7) gives, after re-arrangement:

$$\frac{d[HBr]}{dt} = 2k_2 K_{Br}{}^{1/2} \frac{[Br_2]^{1/2}[H_2]}{1 + k_5[HBr]/k_3[Br_2]} \qquad (6.5.8)$$

Comparison with the experimental rate equation (6.5.1), shows that the mechanism gives the correct dependence on reactant and product concentrations. This is valuable support for the mechanism, but it does not itself provide definite proof. However, independent kinetic evidence accumulating during the last fifty years has given convincing support.

Rate constants and activation energies of elementary reactions

Comparison of the experimental and theoretical rate equations shows:

$$k = 2k_2 K_{Br}^{1/2} \tag{6.5.9}$$

$$k_i = k_5 / k_3 \tag{6.5.10}$$

Accurate values of K_{Br} can be calculated by statistical mechanics or thermodynamics and equation (6.5.9) shows that measuring k gives the elementary rate constant k_2. Reaction (6.5.V) is the reverse of (6.5.II), so independently calculating the equilibrium constant $K = k_2/k_5$ results in a value for k_5. Finally, equation (6.5.10) shows that k_5 and the experimental value of k_i give elementary constant k_3.

For each of the rate constants calculated in this way, the activation energy can be obtained from the 'overall' activation energies found experimentally for k and k_i.

Photochemical reaction

Photochemical studies, in addition to their intrinsic interest, provide further support for the mechanism and allow deduction of the remaining rate constants k_1 and k_4. The experimental rate equation is[12]

$$\frac{d[HBr]}{dt} = \frac{k' I_a^{1/2}[H_2]}{[M]^{1/2}(1 + k_i[HBr]/[Br_2])} \tag{6.5.11}$$

where I_a is the intensity of radiation absorbed, k' is the new experimental rate constant and k_i has the same value found in the thermal system. The mechanism should be identical with that above except that thermal initiation is replaced by photochemical initiation:

$$Br_2 + h\nu \rightarrow 2Br$$

the specific rate of which is $d[Br]/dt = 2I_a$. The steady-state calculation gives:

$$\frac{d[HBr]}{dt} = \frac{2k_2 I_a^{1/2}[H_2]}{k_4^{1/2}[M]^{1/2}(1 + k_5[HBr]/k_3[Br_2])} \tag{6.5.12}$$

This again has the correct form and the $[M]^{-1/2}$ term confirms the termolecular nature of (6.5.IV). Comparing (6.5.11) and (6.5.12) shows $k' = 2k_2/k_4^{1/2}$ and with k_2 known from the thermal system this gives k_4. But $K_{Br} = k_1/k_4$, and calculation of K_{Br} allows deduction of k_1.

The quantum yield (ϕ) is readily measured in the photochemical system:

$$\phi = \frac{\text{number of HBr molecules formed}}{\text{number of quanta absorbed}}$$

Inspecting the mechanism shows that each quantum absorbed provides two bromine atoms, and one bromine atom yields two HBr molecules per 'chain cycle'. Thus

$$\phi = \frac{2(\text{number of chain cycles})}{\frac{1}{2}(\text{number of Br atoms produced})}$$

[12] M. Bodenstein and H. Lütkemeyer, *Z. Phys. Chem.*, **114**, 208 (1925).

and since

$$\frac{\text{number of chain cycles}}{\text{number of Br atoms formed}} = \text{chain length}$$

$$\text{chain length} = 4\phi.$$

At 500 K and 100 Torr, $1{\cdot}3 \times 10^4$ N m^{-2}, reactant pressure, the chain length is about 100.

6.6 Other Hydrogen–Halogen Reactions

The hydrogen–chlorine reaction is fast, not inhibited by HCl product, with a chain length of about 10^6.[13] As with hydrogen–bromine, dissociation of the halogen provides initiation and halogen atom recombination terminates the chain. The essential difference is that the propagation step:

$$Cl^{\cdot} + H_2 \rightarrow HCl + H^{\cdot}$$

analogous to (6.5.II), is only 5 kJ mol^{-1} endothermic and much faster than (6.5.II). Similarly, its reverse reaction is not so fast as (6.5.V) and does not compete successfully for H atoms. Hence chain propagation is fast, with no significant inhibition.

The hydrogen–iodine reaction is slow, largely non-chain. The propagation step

$$I^{\cdot} + H_2 \rightarrow HI + H^{\cdot}$$

is very endothermic with activation energy 140 kJ mol^{-1} and so slow that it, and the chain reaction, predominate only above 800 K. The non-chain reaction is discussed in section 6.2. The hydrogen–fluorine reaction, due to the very strong HF bond and very weak FF bond, is by far the most exothermic of the series. It is very difficult to study, being fast even at liquid hydrogen temperatures and strongly influenced by the vessel surface. Initiation is heterogeneous, both propagation steps are exothermic, and the HF formed in $H + F_2 \rightarrow HF^{\dagger} + F$ contains sufficient energy to dissociate fluorine on collision[14]. Thus a form of energy branching occurs (see sections 6.4 and 7.5).

6.7 The Pyrolysis of Hydrocarbons: Ethane

Hydrocarbon pyrolyses are important both industrially and in the history and development of gas kinetics. From its early stages the petrochemical industry has supplied hydrocarbons as a major basis for the chemical industry and as fuel and power sources for society. One of the main industrial processes involves the conversion of heavy alkanes (paraffins) into lighter, unsaturated molecules by heating, often in the presence of catalyst. The pyrolysis of hydrocarbons gives thermal 'cracking' yielding such products as butadiene, butene, propylene and ethylene. Ethylene is one of the most important reagents used by the chemical industry as a building block in syntheses. Its polymers, the polythenes, form the basis for films, moulding, and pipes, but ethylene

[13] M. Bodenstein, *Z. Phys. Chem.*, **85**, 329 (1931).
[14] J. R. Levy and B. K. W. Copeland, *J. Phys. Chem.*, **72**, 3168 (1968).

may also be converted to ethylene glycol from which polyester fibres and some detergents are derived, and to styrene for polystyrene plastics and synthetic rubbers. A major source of ethylene, particularly in the USA, is the pyrolysis of ethane derived from natural gas. Though this and other industrial processes are usually carried out catalytically today, understanding the homogeneous gas reactions is clearly important.

When ethane is pyrolysed at temperatures of 700–900 K and pressures above 100 Torr, $1{\cdot}3 \times 10^4\ \mathrm{N\ m^{-2}}$, merely explaining the variety of products presents quite a challenge, even before detailed kinetics are considered. In the early stages of the reaction about 99% of the reaction products can be represented by the overall stoichiometry:

$$C_2H_6 \rightarrow C_2H_4 + H_2$$

In addition, much smaller yields of methane and butane are obtained and as the reaction proceeds there is a steady growth in the yield of methane and propylene. This discussion will concentrate on the early part of the reaction which is best understood, and where the kinetics show first-order dependence on ethane concentration

$$-\mathrm{d}[C_2H_6]/\mathrm{d}t = k_{\mathrm{exp}}[C_2H_6] \qquad (6.7.1)$$

Inhibition

For some years many kineticists considered that inhibition studies—adding radical scavengers to the system—provided useful insights into the reaction mechanism. Inhibition did not result in complete suppression of reaction but the rate fell to a low residual value, thought by many to represent the influence of molecular rather than radical reactions. However, as Purnell states[15]: 'for a long period the issues were totally confused by the results of inhibition experiments, notably those involving addition of nitric oxide or propylene. The findings were widely, though not everywhere taken to prove the occurrence of concurrent molecular and radical chain routes to decomposition.' However, 'that there is, in fact, no significant molecular component in paraffin pyrolyses took long to establish but cannot now be questioned.' On this basis a large slice of kinetics history will be set aside to focus on uninhibited alkane pyrolysis which is sufficiently complicated in itself.

Mechanism (Rice–Herzfeld)

Rice and Herzfeld[16] postulated a general mechanism whose main features should be applicable to all hydrocarbon pyrolyses. Taking the ethane system as specific example this gives:

Initiation: the molecule splits at its weakest bond, C—C for ethane

$$C_2H_6 \xrightarrow{k_1} 2CH_3^{\cdot} \qquad (6.7.\mathrm{I})$$

Propagation: (a) a radical formed in the initiation step abstracts a hydrogen atom from the parent

$$CH_3^{\cdot} + C_2H_6 \xrightarrow{k_2} CH_4 + C_2H_5^{\cdot} \qquad (6.7.\mathrm{II})$$

[15] J. H. Purnell and D. A. Lethard, *Ann. Rev. Phys. Chem.*, **21**, 197 (1970).
[16] F. O. Rice and K. F. Herzfeld, *J. Am. Chem. Soc.*, **56**, 284 (1934); K. J. Laidler and B. W. Wojciechowski, *Proc. R. Soc. (London)*, **A260**, 91 (1961).

(b) the new radical undergoes unimolecular dissociation to a stable olefin

$$C_2H_5^{\cdot} \xrightarrow{k_3} C_2H_4 + H^{\cdot} \qquad (6.7.III)$$

Where hydrogen atoms are generated this is followed by

$$H^{\cdot} + C_2H_6 \xrightarrow{k_4} H_2 + C_2H_5^{\cdot} \qquad (6.7.IV)$$

Reactions (6.7.III) and (6.7.IV) form a chain system with $C_2H_5^{\cdot}$ and $H^{\cdot}$ as chain carriers. Some authors describe reaction (6.7.II), in which $C_2H_5^{\cdot}$ is formed, as an extra initiation step.

Termination: in most studies the reactant pressures were above the range where heterogeneous termination is important, thus homogeneous radical combination predominates. This raises the question: which radical of the several present is most probably involved in termination? The abstraction reactions in which $CH_3^{\cdot}$ and $H^{\cdot}$ are removed, (6.7.II) and (6.7.IV), are exothermic and fast compared with the unimolecular dissociation of $C_2H_5^{\cdot}$. The $CH_3^{\cdot}$ and $H^{\cdot}$ concentrations will thus be lower than $C_2H_5^{\cdot}$ and it should be this last radical that participates in termination steps. These will be:

$$2C_2H_5^{\cdot} \xrightarrow{k_{5a}} nC_4H_{10} \qquad (6.7.V)$$

$$2C_2H_5^{\cdot} \xrightarrow{k_{5b}} C_2H_4 + C_2H_6 \qquad (6.7.VI)$$

The basic mechanism (6.7.I–VI) should explain the product yields and first-order kinetics.

Product yields

The list of products given by the mechanism—methane, ethylene, hydrogen and butane—is correct. Experiment shows that 99% of products in the early stages of the reaction are ethylene and hydrogen in equal part. It immediately follows from the mechanism that the chain length, the number of times (6.7.III) and (6.7.IV) are repeated before termination, is about 100. In addition, initiation and termination give minor yields of methane and butane and a little extra ethylene.

Rate equation

From the mechanism, the rate of removal for ethane is:

$$-d[C_2H_6]/dt = k_1[C_2H_6] + k_2[CH_3][C_2H_6] + k_4[H][C_2H_6] - k_{5b}[C_2H_5]^2 \qquad (6.7.2)$$

To remove radical concentrations from the rate equation, the steady-state hypothesis is applied:

$$d[CH_3]/dt = 2k_1[C_2H_6] - k_2[CH_3][C_2H_6] = 0 \qquad (6.7.3)$$

$$d[C_2H_5]/dt = k_2[CH_3][C_2H_6] - k_3[C_2H_5] + k_4[H][C_2H_6] - 2k_{5a}[C_2H_5]^2 - 2k_{5b}[C_2H_5]^2 = 0 \qquad (6.7.4)$$

$$d[H]/dt = k_3[C_2H_5] - k_4[H][C_2H_6] = 0 \qquad (6.7.5)$$

From these simultaneous equations:

$$k_1[C_2H_6] = (k_{5a} + k_{5b})[C_2H_5]^2 \qquad (6.7.6)$$

that is, initiation and termination balance in the steady state. From (6.7.6):

$$[C_2H_5] = \left(\frac{k_1}{k_{5a} + k_{5b}}[C_2H_6]\right)^{1/2} \qquad (6.7.7)$$

From equations (6.7.2)–(6.7.5), the rate equation can be transformed to:

$$-d[C_2H_6]/dt = 3k_1[C_2H_6] + k_3[C_2H_5] - k_{5b}[C_2H_5]^2 \qquad (6.7.8)$$

which, from (6.7.7) gives:

$$-\frac{d[C_2H_6]}{dt} = \underbrace{\left(3k_1 - \frac{k_{5b}k_1}{k_{5a} + k_{5b}}\right)[C_2H_6]}_{(A)} + \underbrace{k_3\left(\frac{k_1}{k_{5a} + k_{5b}}\right)^{1/2}[C_2H_6]^{1/2}}_{(B)}$$

The right-hand side contains two terms: A, which is first order in ethane, depends on rate constants for initiation and termination steps; B, $\frac{1}{2}$ order in ethane, depends on the rate constants for initiation, termination and propagation. Since it is a long chain, the propagation steps must be faster than initiation and termination and to a good approximation term A may be neglected leaving:

$$-\frac{d[C_2H_6]}{dt} = k_3\left(\frac{k_1}{k_{5a} + k_{5b}}\right)^{1/2}[C_2H_6]^{1/2} \qquad (6.7.9)$$

So, at first sight, the mechanism gives $\frac{1}{2}$-order kinetics not the first-order found experimentally. To resolve this contradiction reaction (6.7.III) must be examined in more detail[17].

(6.7.III) is a unimolecular reaction and, as discussed in sections 5.9–14, its rate is pressure dependent. At high pressures the specific rate of (6.7.III) will be $k_3[C_2H_5]$ and at very low pressures the rate will be $k_3'[C_2H_5][M]$. For small percentage decomposition of ethane, the total pressure is almost entirely due to ethane, $[M] \simeq [C_2H_6]$, and the rate of (6.7.III) will be $k_3'[C_2H_5][C_2H_6]$. Between these pressure extremes will be the 'fall-off' region where kinetic order changes from first to second. It appears that most work on ethane has been carried out in the intermediate range where the rate of (6.7.III) is best represented by $k_3'[C_2H_5][C_2H_6]^{1/2}$. Thus in (6.7.9), k_3 should be replaced by $k_3'[C_2H_6]^{1/2}$ giving

$$-\frac{d[C_2H_6]}{dt} = k_3'\left(\frac{k_1}{k_{5a} + k_{5b}}\right)[C_2H_6] \qquad (6.7.10)$$

This shows the required first-order dependence. At lower pressures the unimolecular step will enter its second-order region, and the overall order should rise to a limiting value of $\frac{3}{2}$, which has in fact been observed[18].

There has been an immense amount of research into the fine details of hydrocarbon pyrolyses[19] but only a few main points will be mentioned here. In the later stages of pyrolysis in ethane, the product yield is characterized by a parallel rise in methane and

[17] J. H. Purnell and C. P. Quinn, *Proc. R. Soc. (London)*, **A270**, 267 (1962).
[18] M. C. Lin and M. H. Back, *Can. J. Chem.*, **45**, 3165 (1967).
[19] For a recent review see J. H. Purnell and D. A. Lethard, *Ann. Rev. Phys. Chem.*, **21**, 197 (1970).

propylene. This is attributed to reactions between ethyl radicals and the accumulating ethylene product:

$$C_2H_5^{\cdot} + C_2H_4 \rightleftharpoons 1\text{-}C_4H_9^{\cdot} \qquad (6.7.VII)$$

$$1\text{-}C_4H_9^{\cdot} + C_2H_4 \rightleftharpoons 1\text{-}C_6H_{13}^{\cdot}$$

The 1-hexyl radical undergoes isomerization by intramolecular shift of an H atom from 1 to 5 position:

$$1\text{-}C_6H_{13}^{\cdot} \overset{1,5}{\rightleftharpoons} 2\text{-}C_6H_{13}^{\cdot}$$

This facilitates the decomposition

$$2\text{-}C_6H_{13}^{\cdot} \rightarrow C_3H_7^{\cdot} + C_3H_6$$

$$C_3H_7^{\cdot} \rightarrow CH_3^{\cdot} + C_2H_4$$

and methyl radicals react rapidly by (6.6.II) to give methane. Hence (6.7.VII) eventually results in equal amounts of propylene and methane.

Heavier hydrocarbons

Rice–Herzfeld mechanisms can be applied to heavier alkanes and other hydrocarbons. The mechanistic details are inevitably more complicated, for example with n-hexane initiation could yield:

$$nC_6H_{14} \rightarrow CH_3^{\cdot} + C_5H_{11}^{\cdot}$$

$$\rightarrow C_2H_5^{\cdot} + C_4H_9^{\cdot}$$

$$\rightarrow 2C_3H_7^{\cdot}$$

There is a corresponding increase in possibilities for propagation steps.

It is probably true to say that work on a wide range of hydrocarbons is reaching the stage of elucidating fine details, with the general features well established.

Rate parameters

Taking independent measurements for the rate constants of initiation and termination steps makes it possible to calculate the rate constants and activation energies for a wide range of propagation reactions, H abstractions and dissociations in all hydrocarbon pyrolyses studied.

6.8 Bond Dissociation Energies

The preceding sections have shown how the activation energies of many elementary radical reactions are deduced from kinetic data for complex reactions and such measurements can provide further useful information.

As illustrated in figure 1.5, for example, the difference in activation energies for forward and back reaction in a given system, E_f and E_b, is the difference between energies of reactants and products—the energy change for reaction, ΔE. This will be very similar in value to the heat of reaction, ΔH, which is an enthalpy change. Often

ΔH is calculated readily from the appropriate bond dissociation energies, D, which are thus most useful quantities for the kineticist and they can be measured in several ways. Values of D are conventionally referred to 298 K, but measurement at other temperatures will not usually result in error greater than the experimental error of the measurement. For certain diatomic molecules very accurate D values can be derived from spectroscopy, but reliable measurements in general are difficult to obtain. Kinetic methods for obtaining D from measured activation energies are available for some classes of reaction, for example:

(i) D from molecular dissociation reactions.

For reactions such as $AB \underset{b}{\overset{f}{\rightleftharpoons}} A^{\cdot} + B^{\cdot}$, if E_f is the activation energy of the forward reaction and E_b that for the back reaction then the A—B bond dissociation energy, $D(A—B)$, is given by: $D(A—B) = E_f - E_b + RT$, the RT arising because D is an enthalpy change. However RT is a small term, about 2·5 kJ mol^{-1} at 298 K, and may be neglected. The reverse reaction is a radical combination reaction with approximately zero activation energy (see table 6.2) hence $D(A—B) = E_f$.

(ii) D from hydrogen abstraction by halogen atoms[20].

For reactions of the type: $X^{\cdot} + R—\mathrm{H} \underset{b}{\overset{f}{\rightleftharpoons}} \mathrm{H}X + R^{\cdot}$, then: $D(R—\mathrm{H}) - D(\mathrm{H}—X) = E_f - E_b$. The bond energies of diatomic hydrogen halides, HX, are well known, so if both activation energies are determined this gives $D(R—\mathrm{H})$. Usually this is a C—H bond in a hydrocarbon or substituted hydrocarbon. Many determinations for E_f have been carried out with thermally or photochemically generated halogen atoms, and the reverse reaction can be studied in systems to which excess HX has been added.

As available data on bond dissociation energies accumulate and can be added to the enormous compilations of heats of formation from thermochemistry it becomes increasingly likely that a required dissociation energy can be obtained from such tabulations by simple thermochemical calculation.

Further Reading

N. N. Semenov, *Some Problems in Chemical Kinetics and Reactivity*, Princeton U.P., 1959 and Pergamon, 1958.

S. W. Benson, *Foundations of Chemical Kinetics*, McGraw-Hill, 1960.

P. G. Ashmore, F. S. Dainton and T. M. Sugden (Eds), *Photochemistry and Reaction Kinetics*, Cambridge U.P., 1967.

F. S. Dainton, *Chain Reactions*, Methuen, 1966.

F. G. R. Gimblett, *Kinetics of Chemical Chain Reactions*, McGraw-Hill, 1970.

J. H. Purnell and D. A. Leathard, *Ann. Rev. Phys. Chem.*, **21**, 197 (1970).

[20] G. B. Kistiakowsky and E. R. Van Artsdalen, *J. Chem. Phys.*, **12**, 469 (1944); O. M. Golden and S. W. Benson, *Chem. Rev.*, **69**, 125 (1969).

Exercises

6.1
Methane and ethane are among the products formed when acetone is photolysed. The mechanism is:

$$CH_3COCH_3 + h\nu \rightarrow CH_3CO^{\cdot} + CH_3^{\cdot}$$

$$CH_3CO^{\cdot} \rightarrow CO + CH_3^{\cdot}$$

1 $\quad CH_3^{\cdot} + CH_3COCH_3 \rightarrow CH_4 + C^{\cdot}H_2COCH_3$

2 $\quad 2CH_3^{\cdot} \rightarrow C_2H_6$

If a hydrocarbon, RH, that does not absorb the radiation is added, there is a further source of methane

3 $\quad CH_3^{\cdot} + RH \rightarrow CH_4 + R^{\cdot}$

In a series of experiments at constant temperature and constant initial concentration of acetone, the variation of the ratio $R_M / R_E^{1/2}$ with RH concentration was studied. R_M and R_E are the rates of methane and ethane formation which were measured at very small per cent photolysis. A plot of $R_M / R_E^{1/2}$ versus $[RH]$ gave a straight line graph of slope S. The experiments were repeated at a series of temperatures with the results shown in the table:

T/K	475	500	525	550	575
$S/\mathrm{mol^{-1/2}\,cm^{3/2}\,s^{-1/2}}$	1·35	2·29	3·70	5·73	8·55

The rate constant for reaction 2 is $2 \times 10^{13}\ \mathrm{mol^{-1}\,cm^3\,s^{-1}}$ and its activation energy is zero. Evaluate the Arrhenius parameters for reaction 3.

(Ans: $k_3 = 2{\cdot}5 \times 10^{11}\ \mathrm{e}^{-E_3/RT}\ \mathrm{mol^{-1}\,cm^3\,s^{-1}}$; $E_3 = 42\ \mathrm{kJ\,mol^{-1}}$)

6.2
The photochemical reaction between hydrogen and iodine at 500 K is believed to proceed by a radical non-chain mechanism:

1 $\quad I_2 + h\nu \rightarrow 2I^{\cdot}$

2 $\quad 2I^{\cdot} + I_2 \rightarrow I_2 + I_2$

3 $\quad 2I^{\cdot} + H_2 \rightarrow I_2 + H_2$

4 $\quad 2I^{\cdot} + H_2 \rightarrow 2HI$

The reaction is slow because only a small fraction of termolecular collisions between iodine atoms and a hydrogen molecule result in the formation of hydrogen iodide ($k_3 > k_4$).

In a series of experiments at different $[I_2]/[H_2]$ ratios, the ratio $I_a/(d[HI]/dt)$, where I_a is the intensity of light absorbed, was measured with the results shown in the table

$\{I_a/(d[HI]/dt)\}$	7·14	11·3	15·5	19·7	23·9
$([I_2]/[H_2])/10^{-2}$	1	3	5	7	9

Find the relative efficiencies of hydrogen and iodine molecules as third bodies for iodine atom recombination and the ratio k_4/k_3.

(Ans: Relative efficiency, $k_2/k_3 = 46$; $k_4/k_3 = 0{\cdot}11$)

6.3
The following rate equation was obtained for the reaction between hydrogen and chlorine in the presence of oxygen:

$$d[HCl]/dt = k[H_2][Cl_2]^2/[O_2]([H_2] + k'[Cl_2])$$

The proposed mechanism is:

1	$Cl_2 + M$	$\rightarrow 2Cl^\cdot + M$
2	$Cl^\cdot + H_2$	$\rightarrow HCl + H^\cdot$
3	$H^\cdot + Cl_2$	$\rightarrow HCl + Cl^\cdot$
4	$H^\cdot + O_2 + M$	$\rightarrow HO_2^\cdot + M$ ($HO_2 \rightarrow$ wall)
5	$Cl^\cdot + O_2 + M$	$\rightarrow ClO_2^\cdot + M$ ($ClO_2 \rightarrow$ wall)

Under the experimental conditions (a) for the rate of formation of [HCl], $k_3[H][Cl_2] \gg k_2[Cl][H_2]$; (b) the term $k_4k_5[O_2]^2[M]^2$ may be neglected. Show that the mechanism is consistent with the rate equation and that $k = 2k_1k_3/k_4$ and $k' = k_3k_5/k_2k_4$.

6.4
The experimental rate equation for the pyrolysis of acetaldehyde is $-d[CH_3CHO]/dt = k[CH_3CHO]^{3/2}$.

A simple Rice–Herzfeld mechanism for the reaction is

1	CH_3CHO	$\rightarrow CH_3^\cdot + CHO^\cdot$
2	$CHO^\cdot$	$\rightarrow CO + H^\cdot$
3	$H^\cdot + CH_3CHO$	$\rightarrow CH_3CO^\cdot + H_2$
4	$CH_3^\cdot + CH_3CHO$	$\rightarrow CH_4 + CH_3CO^\cdot$
5	$CH_3CO^\cdot$	$\rightarrow CH_3^\cdot + CO$
6	$2CH_3^\cdot$	$\rightarrow C_2H_6$

Show that this mechanism is consistent with the rate equation if $k_4(k_1/k_6)^{1/2} \gg k_1$. Compare the kinetics of this reaction with ethane pyrolysis (section 6.7) with particular reference to the effect of pressure on reaction order.

6.5
The free radical polymerization of an olefin X proceeds by the following mechanism in which the details of initiation are not known.

Initiation	$\rightarrow R^\cdot$
$R^\cdot + X$	$\rightarrow R_1^\cdot$

$$R_1^{\cdot} + X \rightarrow R_2^{\cdot}$$

$$\vdots$$

$$R_n^{\cdot} + X \rightarrow R_{n+1}^{\cdot}$$

$$R_n^{\cdot} + R_m^{\cdot} \rightarrow \text{products}$$

A simple kinetic treatment is obtained if it is assumed that all chain propagation steps have the same rate constant, k, and all termination steps have the same rate constant, k'. If θ is the rate of initiation and $\Sigma[R_n^{\cdot}]$ is the total radical concentration, show that:

$$\theta = k'(\sum[R_n^{\cdot}])^2 \qquad \text{and} \qquad -\mathrm{d}[X]/\mathrm{d}t = k(\theta/k')^{1/2}[X]$$

6.6

In a chain reaction, radical X is formed in a homogeneous initiation step of rate θ and is removed in a homogeneous termination step that is first order in radical concentration and of rate $g[X]$. Show that if θ and g remain approximately constant, the radical concentration reaches a steady value θ/g when time, $t \gg 1/g$.

6.7

The gas phase reaction $A_2 + B_2 \rightarrow 2AB$ was investigated at temperatures in the 500–600 K range in a vessel whose surface had been treated to ensure that heterogeneous reactions were unimportant. The rate of product formation fitted the equation:

$$\mathrm{d}[AB]/\mathrm{d}t = k_{\mathrm{exp}}[A_2][B_2]^{1/2}$$

From spectroscopic data it is known that the bond dissociation energies for the diatomic molecules A_2, B_2 and AB are 400, 100 and 400 kJ mol^{-1} respectively. Write a mechanism that can be shown, using the stationary-state hypothesis, to be consistent with the rate equation. The measured overall activation energy E_{exp} was 155 kJ mol^{-1}. Estimate the activation energies of the elementary reactions.

(Ans: The mechanism, with approximate activation energies in kJ mol^{-1}, is: (i) initiation, dissociation of B_2, $E \sim 100$; (ii) propagation reaction of B, $E \sim 105$; (iii) propagation reaction of A, $E \sim 0$; (iv) termination, combination of B, $E \sim 0$)

Chapter 7: Branched-Chain Reactions

Branched-chain reactions are absorbingly interesting in theory because of the extraordinary kinetic consequences of one extra type of elementary step, and in practice due to the economic importance of the branched-chain combustion processes. Combustion essentially involves two reactants, a *fuel* and an *oxidant*, for example a hydrocarbon and oxygen respectively. It is a self-sustaining reaction that sometimes proceeds at a steady rate but beyond certain critical experimental conditions the reaction becomes *autocatalytic* and the rate undergoes *auto-acceleration* leading to explosion. The onset of auto-acceleration is also called *ignition*, and it usually starts in a localized volume, explosive reaction then spreading rapidly throughout the vessel. If the vessel is a long tube and reaction is initiated at one end a *combustion wave* travels down the tube, and if the reaction zone is luminous it is called a *flame*. Using suitable burners a stationary flame can be set up—as in the bunsen burner or domestic gas cooker. In some systems a combustion wave travels faster than sound, becoming a *detonation wave* (a shock wave). Finally, combustion can be generated in certain liquids or solids chosen to yield maximum energy in minimum time—to act as *explosives*.

Combustion is employed widely in heating appliances, internal combustion engines, other propellants and high explosives. Such practical systems are complicated by the engineering of the 'reaction vessel' and the fact that fuels and oxidants often contain many components—a petrochemical hydrocarbon mixture as fuel and air as oxidant, for example[1].

Faced with such complexities it is vital to grasp the basic kinetic principles governing such behaviour. These are illustrated here for the hydrogen–oxygen and carbon

[1] For a general account of combustion systems see J. N. Bradley, *Flame and Combustion Phenomena*, Science Paperbacks, Chapman and Hall, London, 1972.

monoxide–oxygen reactions, and an account is given of hydrocarbon combustion and some associated environmental problems. We start with a discussion of some general mechanistic topics.

7.1 Thermal and Isothermal Explosion

Elementary branching reactions are of crucial importance but explosion is sometimes observed in unbranched reactions. A *thermal explosion* occurs when, in an exothermic reaction, the rate at which heat is evolved exceeds the rate of heat loss by conduction, convection and radiation. The autocatalytic action required for explosion arises from this self-heating because the reaction rates increase dramatically with temperature. A simple quantitative account was given by Semenov[2] along the following lines. If the temperature of the reactant gas is T and the vessel walls and surroundings are at a lower temperature, T_s, the rate of heat loss is given

$$Q_- = KS(T - T_s)$$

where K is a coefficient of heat transfer, and S the surface area of the reaction vessel.

In a reaction with exothermicity ΔH and rate R, in vessel of volume V the rate of heat production is given

$$Q_+ = R\,\Delta HV$$

Assuming that the reaction rate is approximately kC^n where C represents reactant concentrations and k is the overall rate constant, and if k can be expressed by Arrhenius parameters $k = A\,\mathrm{e}^{-E/RT}$, then

$$Q_+ = A\,\mathrm{e}^{-E/RT}C^n\,\Delta HV$$

Figure 7.1 shows the variation of Q_- and Q_+ with temperature. For a given vessel there is a single linear plot for Q_-, but Q_+ depends on reactant concentrations and possible curves are drawn in figure 7.1 for different initial concentrations, $C_{III} > C_{II} > C_I$.

The criterion for autocatalysed thermal explosion is that $Q_+ > Q_-$ throughout the reaction. This is the case for mixture III at any initial temperature. With mixture I explosion will result if the initial temperature is T_{ign} or above, but if the initial temperature is below T_{st} the reaction rate increases until T_{st} is reached. Then the rates of heat gain and loss are identical and a steady rate is maintained. This state is self-stabilizing since if the temperature rises slightly heat loss exceeds heat gain and temperature falls to T_{st}. Similarly if the initial temperature is between T_{ign} and T_{st} it falls to T_{st}. Mixture II is the critical condition where T_{ign} and T_{st} coincide at T_c, so this mixture represents the lowest possible concentration for spontaneous explosion and is at an *explosion limit*.

In contrast to unbranched systems of this type, in reactions where the fundamental reason for accelerating rate is the increase of radical concentration in branching reactions explosion can occur, in theory, without this thermal effect. Such explosions

[2] N. N. Semenov, *Some Problems in Chemical Kinetics and Reactivity*, Vol. 2, Pergamon, Oxford, 1958.

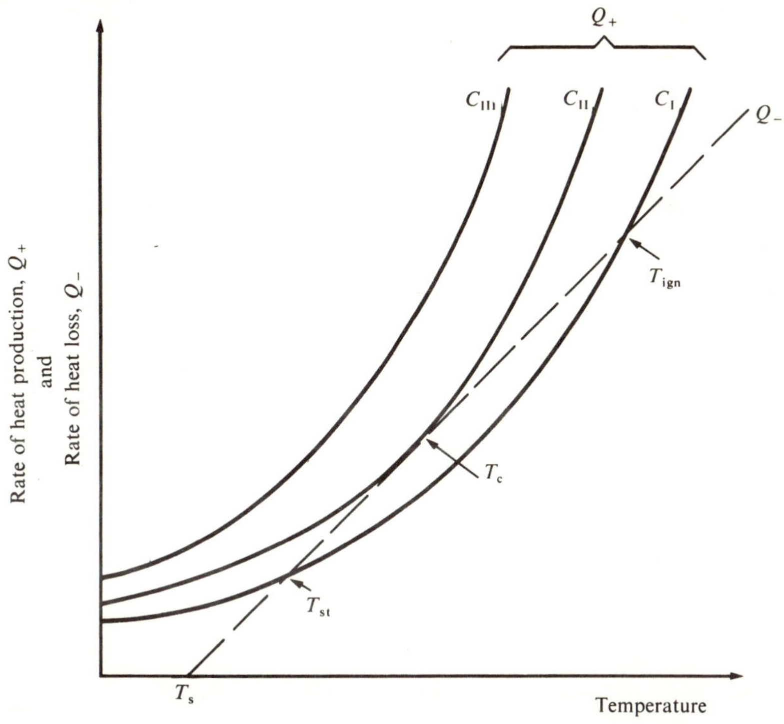

Figure 7.1
Rate of heat production, Q_+, versus temperature for three different concentrations C_{I}, C_{II}, C_{III} of reactants, and, rate of heat loss, Q_-, versus temperature.

are termed *isothermal*. In practice branched-chain reactions are themselves exothermic overall and the acceleration in reaction rate is inevitably enhanced by a thermal contribution.

7.2 The Branching Factor

In systems where more than one radical is involved in the chain, discussion can be simplified by noting that to a very good approximation these radicals rapidly reach a steady ratio of concentrations and it is valid to refer to a generalized radical concentration $[\bar{x}]$. Radical concentrations do not change in elementary chain propagation steps, only in initiation, branching and termination steps.

For each reaction type, the rate of change for radical concentration is given by multiplying the radical concentrations by rate coefficients, which are combinations of rate constants and reactant concentrations:

$$\mathrm{d}[\bar{x}]/\mathrm{d}t = R_{\mathrm{i}} + f[\bar{x}] - g[\bar{x}] + f'[\bar{x}]^2 - g'[\bar{x}]^2 \tag{7.2.1}$$

The rate coefficients are: R_i for initiation; f and f' for branching reactions which are first and second order (linear and quadratic) in radical concentration; and g and g' for linear and quadratic radical termination steps.

Equation (7.2.1) may be rewritten:

$$d[\bar{x}]/dt = R_i + \phi[\bar{x}] + \phi'[\bar{x}]^2 \qquad (7.2.2)$$

where $\phi = f - g$, the difference in rate coefficients for linear branching and termination. It is called the *linear branching factor*. Similarly $\phi' = f' - g'$ is the *quadratic branching factor*†. The kinetic details of any reaction depend critically on the particular combination of linear and quadratic branching and termination steps. Important kinetic consequences can be illustrated for a relatively simple case which does, however, correspond to important practical systems—a chain reaction linearly branched and linearly terminated. In this case the rate of change in radical concentration is:

$$d[\bar{x}]/dt = R_i + \phi[\bar{x}] \qquad (7.2.3)$$

Integration of (7.2.3), with $[\bar{x}] = 0$ at $t = 0$ gives two possible results. If ϕ is positive, $f > g$:

$$[\bar{x}] = R_i(e^{\phi t} - 1)/\phi \qquad (7.2.4)$$

If ϕ is negative, $g > f$, then putting its value $-|\phi|$:

$$[\bar{x}] = R_i(1 - e^{-|\phi|t})/|\phi| \qquad (7.2.5)$$

The results expressed in (7.2.4) and (7.2.5) are plotted in figure 7.2, where the crucial difference is evident. Since the overall reaction rate is proportional to radical concentration, these also give the form of the rate versus time curves. The diagram shows that, if the termination rate exceeds the branching rate ($\phi < 0$), the result is a steady radical concentration and reaction rate, experimental features similar to those of unbranched-chain reactions. If, however, branching exceeds termination ($\phi > 0$) continuously accelerating reaction leads to explosion. Thus autocatalytic behaviour is clearly associated with the branching step. The critical condition $\phi = 0$ means that minute changes in reactant concentrations or experimental conditions can give ϕ either slightly positive, leading to explosion, or slightly negative, which maintains the steady reaction. So $\phi = 0$ is the criterion for the *explosion limit.*

Induction period

In practice the onset of explosion is detected through some phenomenon such as sudden temperature rise or visible light emission. Whatever property is observed, detection will require some minimum radical concentration. The time at which this detectable concentration is reached may vary with reaction conditions, and is referred to as the *induction period.* If $[\bar{x}]_i$ is the critical concentration denoting the end of the induction period at time t_i, then from (7.2.4)

$$t_i = \frac{1}{\phi}\ln\left(1 + \phi\frac{[\bar{x}]_i}{R_i}\right) \qquad (7.2.6)$$

† Using the same symbol for branching factor and quantum yield is an unfortunate convention.

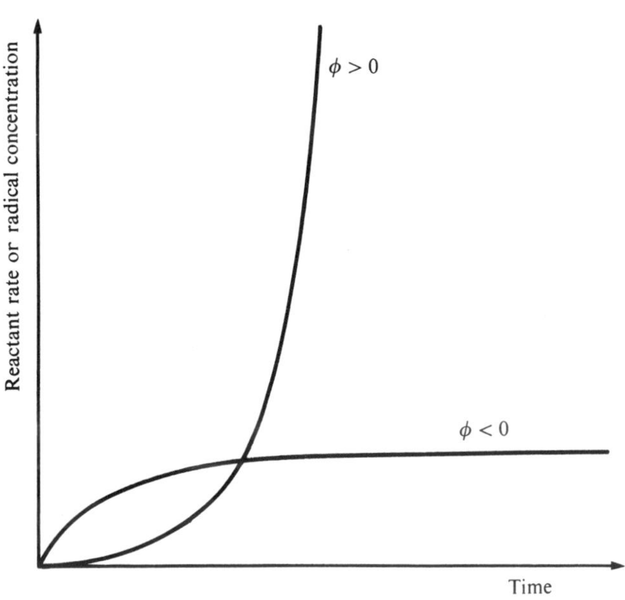

Figure 7.2
Variation of general radical concentration (or of reaction rate) with time for chain reactions with positive and negative linear branching factors.

Inspection of equation (7.2.6) shows that with large positive branching factors the induction period will be very short, but if reactant pressures are such that ϕ only just exceeds 0, long induction periods occur.

This general discussion reveals that some of the experimental features described in section 6.3 are consequences of the elementary branching reactions, and particularly of the competition between branching and termination. These points will now be illustrated for some specific combustion reactions. It should be emphasized that this account has been limited to linear branching and termination. Important differences arise if, for example, termination is quadratic. Then $\phi > 0$ is no longer a *sufficient* criterion for explosion.

7.3 The Hydrogen–Oxygen Reaction: Experimental

This is the most intensively studied reaction in kinetics[3], both for its intrinsic interest as a branched-chain reaction and as a prototype of the economically important combustion reactions. In fact hydrogen may itself become a commercially useful fuel in the future.

[3] A more detailed account of the hydrogen–oxygen reaction is given by R. R. Baldwin and R. W. Walker, *Essays in Chemistry*, Vol. 3, (Eds J. N. Bradley *et al.*), Academic Press, 1972.

The overall reaction, of stoichiometry:

$$2H_2 + O_2 \rightarrow 2H_2O$$

can be explosive at temperatures above 700 K. For many years researchers obtained inconsistent results, erratic behaviour that is probably associated with the influence of the vessel surface. It is now usual to use either surfaces such as potassium chloride, which is very efficient at absorbing radicals and promoting surface combination, or surfaces of opposite character—such as boric oxide which is highly inert towards radicals. Typical explosion limits for potassium chloride coated vessels are shown in figure 7.3 where reactant pressure refers to stoichiometric mixtures ($2H_2 + O_2$). Figure 7.3 portrays three limits, the first and second defining a peninsular explosion region, the third occurring at very high pressure and undoubtedly having a large thermal contribution. The limits are also affected in different ways by the size (diameter) of the reaction vessel and addition of inert gases. Another way of looking at the limits is to consider how reaction rate changes with reactant pressure, at constant

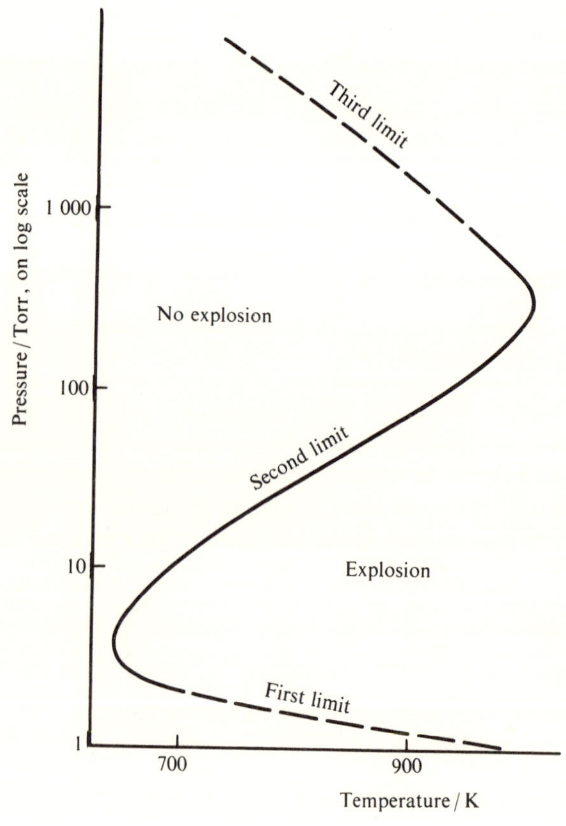

Figure 7.3
Explosion diagram for the H_2–O_2 reaction. Pressure (on log scale) of stoichiometric mixture ($2H_2 + O_2$) versus temperature in KCl coated vessel.

temperature. This is shown in figure 7.4, and the effect of vessel diameter and inert gases is also indicated.

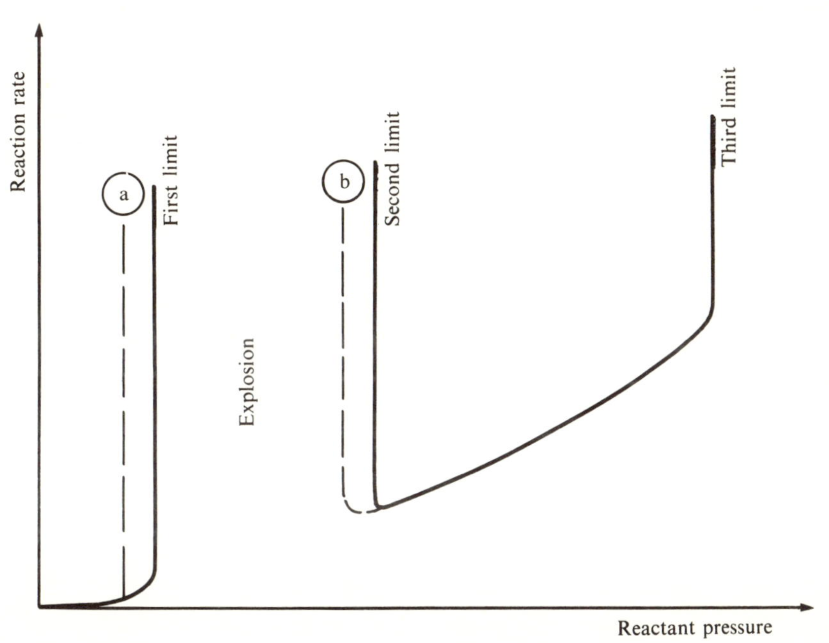

Figure 7.4
Variation of reaction rate with reactant pressure, at constant temperature, for reaction with three explosion limits. The broken lines show the displacement of: (a) first limit on adding inert gas or increasing vessel diameter; (b) second limit on adding inert gas.

Kinetics in the non-explosive region have been thoroughly investigated over many years, but this section will concentrate on explosive behaviour. The mechanism that applies at the explosion limits accounts for behaviour in the non-explosive region with only slight modification. This mechanism must explain the diverse experimental features at the limits which are illustrated in figures 7.3 and 7.4 and may be summarized:

(i) The *first limit* occurs at pressures below about 4 Torr, 5×10^2 N m^{-2},

(a) At constant reactant pressure the limit is reached by increasing temperature.

(b) At constant temperature the limit is reached on increasing reactant pressure.

(c) Increasing the diameter of the reaction vessel lowers the limit.

(d) Adding inert gas lowers the limit.

It should be emphasized that lowering the first limit extends the explosion region. The effect of temperature, reactant pressure (P_I) and vessel diameter (d) at the first limit is described quantitatively by:

$$P_\mathrm{I} = \text{constant} \times \frac{e^{E_\mathrm{I}/RT}}{d} \tag{7.3.1}$$

(ii) The *second limit* occurs at reactant pressures in the range 4–400 Torr, 5×10^2–5×10^4 N m^{-2},

(a) At constant reactant pressure the limit is reached on increasing temperature.

(b) At constant temperature the limit is reached by *decreasing* reactant pressure.

(c) The limit is not significantly affected by varying vessel diameter.

(d) Adding inert gas lowers the limit, that is it diminishes the explosion region.

The temperature and pressure effects (and the absence of vessel diameter's influence) are described quantitatively by:

$$P_{II} = \text{constant} \times e^{-E_{II}/RT}$$

(iii) The *third limit*. At such pressures and temperatures reaction rates are very fast even in the steady region, and this limit is difficult to study quantitatively. Qualitatively the most important observation is that the limit is reached on increasing reactant pressure.

7.4 Hydrogen–Oxygen Reaction: Mechanism

These experimental characteristics are of such diversity it is gratifying that a quite simple mechanism gives successful qualitative and semi-quantitative account.

(i) First limit

At such low pressures, the experimental evidence suggests a mechanism governed by competition between heterogeneous termination and a homogeneous branching step of higher order.

Initiation	$H_2 + O_2 \xrightarrow{\text{wall}} 2OH^{\cdot}$	(7.4.I)
Propagation	$OH^{\cdot} + H_2 \xrightarrow{k_2} H_2O + H^{\cdot}$	(7.4.II)
Branching	$H^{\cdot} + O_2 \xrightarrow{k_3} OH^{\cdot} + O^{\cdot\cdot}$	(7.4.III)
	$O^{\cdot\cdot} + H_2 \xrightarrow{k_4} OH^{\cdot} + H^{\cdot}$	(7.4.IV)
Termination	$H^{\cdot} \xrightarrow{k_5} \text{wall}$	(7.4.V)

The precise nature of the initiation step is uncertain, heterogeneous interaction of hydrogen and oxygen to yield hydroxyl radicals being most likely. Reaction (7.4.II) is a fast exothermic propagation step but branching reaction (7.4.III) is endothermic, of high activation energy, and only fast enough to allow explosion at temperatures above 700 K. Reaction (7.4.III) is followed by the much faster (7.4.IV) and termination is attributed to diffusion of hydrogen atoms to the wall followed by heterogeneous combination. The explosion limits will be governed by competition for H atoms between this termination step and branching reaction (7.4.III). Both are linear (first order in radical concentration) and, as shown in section 7.2, the branching factor is the difference in rate coefficients for branching and termination, $\phi = f - g$. Since the rates of branching and termination are effectively $2k_3[H][O_2]$ and $k_5[H]$ respectively, the appropriate rate coefficients are $f = 2k_3[O_2]$ and $g = k_5$. Thus:

$$\phi = 2k_3[O_2] - k_5$$

The explosion limit is defined by $\phi = 0$ and

$$[O_2] = k_5 / 2k_3 \tag{7.4.1}$$

As described in section 6.4, the rate constant for the diffusion step is given $k_5 = \tau D_0 / d^2 P$. In the stoichiometric reaction mixture the concentration of oxygen is proportional to total pressure, or, $[O_2] = k'P$. Substitution into (7.4.1) gives for the limit at pressure P_I:

$$k'P_I = \frac{\tau D_0}{d^2 P_I 2k_3}$$

or

$$P_I = \left(\frac{\tau D_0}{2k'}\right)^{1/2} \frac{1}{k_3{}^{1/2} d} \tag{7.4.2}$$

k_3 may be expressed in Arrhenius form as $A_3 e^{-E_3/RT}$, the diffusion coefficient D_0 has only a $T^{1/2}$ temperature dependence which may be neglected, and, substituting into equation (7.4.2):

$$P_I = \left(\frac{\tau D_0}{2k' A_3}\right)^{1/2} \frac{e^{E_3/2RT}}{d} \tag{7.4.3}$$

This is equivalent to the experimental relationship for the first limit, equation (7.3.1), and correctly describes variation with temperature, pressure and vessel diameter. Adding inert gas will affect the competition between branching and termination by decreasing the rate of diffusion to the wall, thus favouring explosion and so lowering the first limit.

(ii) Second limit

At the higher pressures of the second limit homogeneous termination should replace the heterogeneous step, and the fact that increasing pressure decreases the probability of explosion (lowers the limit) indicates that the new termination step is of higher order than the branching reaction. The mechanism will be the same as given above except for the replacement of (7.4.V) with the new homogeneous step:

Termination $\quad H^{\cdot} + O_2 + M \rightarrow HO_2{}^{\cdot} + M \qquad$ (7.4.VI)

The HO_2 radical is unusual in that it is relatively unreactive—its probable fate is diffusion to the wall without undergoing reaction in the gas phase*. The explosion limit is now governed by competition for H atoms between reactions (7.4.III) and (7.4.VI). Branching and termination are still linear and with the same kinetic treatment:

$$\phi = 2k_3[O_2] - k_6[O_2][M]$$

At the explosion limit $\phi = 0$, and

$$[M] = 2k_3 / k_6 \tag{7.4.4}$$

In the stoichiometric reaction mixture $[M]$ may be hydrogen or oxygen. More precisely, since hydrogen and oxygen exhibit different efficiencies as third bodies in this

* At the wall, these radicals react relatively slowly to give hydrogen peroxide and oxygen.

termolecular step, their individual concentrations should be multiplied by suitable coefficients and $[M] = \gamma_{H_2}[H_2] + \gamma_{O_2}[O_2]$. However, to a good approximation $[M] = k''P$ and from (7.4.4), for the limit at P_{II}:

$$P_{II} = 2k_3/k''k_6 \tag{7.4.5}$$

The temperature dependences of k_3 and k_6 are given by Arrhenius expressions, and substituting into (7.4.5):

$$P_{II} = \frac{2A_3\,e^{-E_3/RT}}{k''A_6\,e^{-E_6/RT}} = \left(\frac{2A_3}{k''A_6}\right) e^{-(E_3-E_6)/RT} \tag{7.4.6}$$

This has the same form as the experimental relationship for the second limit, equation (7.3.2), and correctly describes variation with temperature and pressure and the absence of effect for vessel diameter. Good quantitative agreement is also obtained comparing experimental E_{II} with $(E_3 - E_6)$. Adding inert gas increases the size of $[M]$ for termination step (7.4.VI) without affecting the rate of branching and so the explosion region is narrowed.

In summary, the simple mechanism suggested above gives a reasonable interpretation of the remarkable and diverse features of the peninsular explosion region.

(iii) Third limit

This can be dealt with qualitatively by attributing the onset of explosion at very high pressures to an extra process which diminishes termination:

$$HO_2^{\cdot} + H_2 \rightarrow H_2O + OH^{\cdot}$$

So HO_2 undergoes a gas phase reaction that competes with its diffusion to the wall and termination.

Non-explosive region

To explain in detail the observed rates in the region between the second and third limits some extra reactions must be postulated. An experimental complication is the detection of hydrogen peroxide, and the full complexity of the system is indicated by the list of reactions that are thought to participate to some extent:

$$2HO_2^{\cdot} \rightarrow H_2O_2 + O_2$$

$$H_2O_2 + M \rightarrow 2OH^{\cdot} + M$$

$$H^{\cdot} + H_2O_2 \rightarrow H_2O + OH^{\cdot}$$

$$H^{\cdot} + H_2O_2 \rightarrow H_2 + HO_2^{\cdot}$$

$$OH^{\cdot} + H_2O_2 \rightarrow H_2O + HO_2^{\cdot}$$

$$H^{\cdot} + HO_2^{\cdot} \rightarrow 2OH^{\cdot}$$

Stationary-state treatment for first and second limits

It is reasonable to apply the stationary-state method to the region outside explosion limits and it is easily shown that the steady concentration of hydrogen atoms is, including both termination steps (7.4.V) and (7.4.VI):

$$[H]_{ss} = 2R_i/(k_5 + k_6[O_2][M] - 2k_3[O_2]) \tag{7.4.7}$$

where R_i is the rate of initiation step (7.4.I).

Although it seems a paradox to apply this method to explosions, the explosion limit can be envisaged as the condition for breakdown of the steady state, with hydrogen atom concentration tending to infinity. Thus the denominator in (7.4.7) tends to zero:

$$k_5 + k_6[O_2][M] - 2k_3[O_2] = 0$$

At the first limit, the k_6 term can be neglected and at the second limit the k_5 term is not significant. This leads to the same results as the previous treatment.

7.5 The Carbon Monoxide–Oxygen Reaction

This reaction has been classed as branched-chain for almost as long as the hydrogen–oxygen reaction, and it has received particular attention because of the unique character of the branching step—an excited molecule rather than a radical being involved. The reaction exhibits a peninsular explosion region with similar properties to the first and second limits with hydrogen–oxygen though some workers found a double peninsula, one lying within the other[4]—the outer marked the onset of reaction accompanied by a blue glow emission, the inner corresponds to true explosion. Interpretation of results is confused by extreme experimental difficulties, especially sensitivity to minute traces of water. It is not certain that a genuine 'dry' reaction has ever been observed and rigorous drying of reactants shifts explosion limits to higher temperatures.

The mechanism proposed for the dry reaction[5] has oxygen atoms in initiation and propagation steps

Initiation $\quad CO + O_2 \rightarrow CO_2 + O^{\cdot\cdot}$ (7.5.I)

Propagation $\quad O^{\cdot\cdot} + CO \rightarrow CO_2{}^*$ (7.5.II)

It is not possible to invoke a satisfactory chain branching step with oxygen atoms and the participation of electronically excited carbon dioxide has been postulated, the formation of a ground state molecule in reaction (7.5.II) being spin forbidden. The branching step is then:

Branching $\quad CO_2{}^* + O_2 \rightarrow CO_2 + 2O^{\cdot\cdot}$ (7.5.III)

and (7.5.III) has been called *energy branching*.

Termination is heterogeneous at low pressures

Termination $\quad O^{\cdot\cdot} \rightarrow \text{wall}$ (7.5.IV)

$CO_2{}^* \rightarrow \text{wall}$ (7.5.V)

and homogeneous at high pressures.

$CO + O^{\cdot\cdot} + M \rightarrow CO_2$ (7.5.VI)

$CO_2{}^* + M \rightarrow CO_2 + M$ (7.5.VII)

[4] D. E. Hoare and A. D. Walsh, *Trans. Faraday Soc.*, **50**, 37 (1954).
[5] For example by A. S. Gordon and R. H. Knipe, *J. Phys. Chem.*, **59**, 1160 (1955).

This reaction scheme can be treated in a manner analogous to that described for the hydrogen–oxygen reaction, as it is another example of linear branching and termination at each limit.

Energy branching is also found in the hydrogen–fluorine reaction (see section 6.6) through the agency of *vibrationally* excited HF molecules.

7.6 Combustion of Hydrocarbons: Experimental

Popular methods for domestic heating are the combustion of natural gas, mainly methane (which in Britain has replaced 'town gas', a carbon monoxide–hydrogen mixture) and fuel oil, a mixture of heavy hydrocarbons. Internal combustion engines burn lighter (288–423 K) hydrocarbon fractions distilled from petroleum. Certain compounds such as aromatics, olefins, branched alkanes and 'anti-knock' agents are added to improve combustion efficiency. These practical combustion systems are further complicated by the use of air as oxidant and by the mechanics of the 'vessels' or burners. There is still an urgent need for more detailed understanding of the basic kinetics for pure hydrocarbon–oxygen combustion in simple laboratory vessels. However the objectives of this section must be not only to describe the salient features of their mechanisms, but also to convey a clear impression of just how complicated they are. This will hopefully re-inforce the idea that there will be no simple solution to the many environmental problems associated with different forms of combustion, and that oversimplifying the issues will be counterproductive in the long run.

The main experimental characteristics of these reactions are:

(i) There may be very long *induction periods* lasting many minutes or even hours. They are affected greatly by sensitizers or inhibitors.

(ii) The induction period is followed by an accelerating increase in reaction rate. At high temperatures and pressures this continues to explosion, but at relatively low temperatures and pressures it reaches a *maximum rate*, and then gradually falls.

(iii) The explosion diagram is very complicated, as shown by the schematic representation in figure 7.5. There are not only explosive and non-explosive regions but also an area of irregular boundary which is marked experimentally by the appearance of one or more transient '*cool flames*'. These exhibit blue luminescence that accompanies a series of pressure and temperature pulses before ignition. The light emission has been identified as chemiluminescence from electronically excited formaldehyde.

(iv) The effect of temperature on reaction rates is unusual. In the low and high temperature ranges reaction rates increase with temperature, but in the intermediate range reaction rate has a *negative temperature coefficient*, rate decreasing with temperature. This is true both of non-explosive reaction and where ignition occurs—the slope of section b–c on the boundary of the explosion region shown in figure 7.5 means explosion is more likely on decreasing temperature.

(v) A *variety of products* is formed, differing in character in the low and high temperature regions. At low temperatures alcohols, aldehydes and ketones predominate,

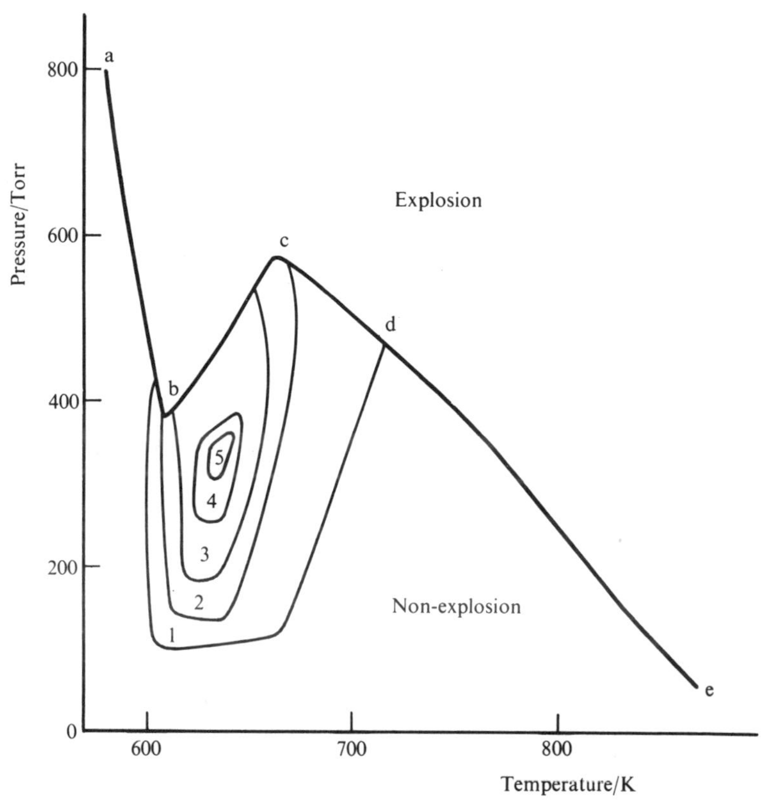

Figure 7.5
Schematic explosion diagram for hydrocarbon–oxygen mixtures over typical pressure and temperature ranges. The cool-flame regions are numbered to indicate how many successive flames are observed.

whereas at high temperatures the major products are unsaturated and saturated hydrocarbons of lower molecular weight than the combustant, together with hydrogen peroxide. The high temperature combustion rather resembles an oxygen catalysed pyrolysis of the hydrocarbon.

(vi) In combustion there is widely *different reactivity* for various hydrocarbons. To note only two trends, reactivity increases with chain length and straight chains are more reactive than branched chains. This is an important factor for efficient performance of the internal combustion engine. If reactivity is too high ignition occurs too rapidly, giving the mechanically undesirable 'knocking'. Ignition may even take place before the initiating spark in the engine (pre-ignition). Thus branched-chain hydrocarbons are preferred as fuel for these engines.

Mechanisms that can explain these experimental features will not be simple. They will be discussed first in general terms, followed by a more detailed application to hydrocarbon systems.

7.7 Combustion of Hydrocarbons: General Mechanistic Features

Induction periods, whether leading to maximum rate or to explosion, are much reduced by sensitizers but they never fall to zero. This indicates a chain reaction that is autocatalytic and the nature of the branching step was characterized by Semenov[6]. Primary reaction results from a straight chain of moderate length, for example, with reactants L, M:

$$\text{Initiation} \rightarrow X^{\cdot}$$

$$X^{\cdot} + L \rightarrow \text{Pr} + Y^{\cdot}$$

$$Y^{\cdot} + M \rightarrow X^{\cdot}$$

$$2X^{\cdot} \rightarrow \text{termination}$$

The product Pr is relatively unstable, though with a lifetime somewhat greater than the lifetime of an average primary chain. Pr may largely decompose to stable molecular products, but a fraction dissociates to radical chain carriers, e.g.

$$\text{Pr} \rightarrow Y^{\cdot} + Z^{\cdot}$$

Such a step is called a *degenerate branching reaction*. If Pr has a lifetime that is short enough to yield new chain carriers before an appreciable fraction of reactants has been consumed in the primary chain, then after a certain time most new chains are initiated by Pr. Hence only after Pr has accumulated for a certain time does the reaction rate accelerate. So this basic mechanism is consistent with autocatalytic reaction and long induction periods.

Chain-thermal explosion

Branched-chain reactions are always exothermic, and even where explosion is referred to as isothermal there is a thermal contribution. For linear branched and terminated chain reactions, it was shown above that the criterion for explosion was a positive branching factor ($\phi > 0$). In systems where termination is mainly by radical–radical combination (quadratic) detailed kinetic analysis shows that $\phi > 0$ is not a sufficient condition for explosion, and purely isothermal ignition does not occur. Hydrocarbon combustions are in this category, and only if the branching factor exceeds some critical value ϕ_c does the branched reaction have a thermal contribution that is high enough for explosion. The explosion is described as *chain-thermal*.

The maximum rate

If experimental conditions are such that the branching factor, though positive, is less than the critical value for chain-thermal explosion a maximum rate is reached. A high proportion of reactants is consumed in the primary chain before the reaction rate can accelerate sufficiently for ignition. ϕ and the overall rate then fall to levels not allowing further acceleration, and a maximum rate results. In contrast where $\phi > \phi_c$ the critical reaction rate for ignition is reached before there is much consumption of reactants.

[6] N. N. Semenov, *Chemical Kinetics and Gas Reactions*, Oxford U.P., 1935.

7.8 Combustion of Hydrocarbons: Detailed Mechanism

The experimental rate data suggest that different mechanisms are operative in the low and high temperature regions and in the intermediate range with negative rate coefficient. Though many details remain obscure the major trends are consistent with the following[7].

Low temperatures

Initiation	$RH + O_2 \rightarrow R^{\cdot} + HO_2^{\cdot}$	(7.8.I)
Propagation	$R^{\cdot} + O_2 \rightarrow ROO^{\cdot}$	(7.8.II)
	$ROO^{\cdot} + RH \rightarrow ROOH + R^{\cdot}$	(7.8.III)
Termination	$R^{\cdot} + R^{\cdot} \rightarrow R_2$	(7.8.IV)

The hydroperoxide product from (7.8.III) can usually be detected at $T < 600$ K. It contains the weak peroxy (O—O) bond and provides degenerate branching followed by another propagation step:

$$ROOH \rightarrow RO^{\cdot} + OH^{\cdot} \tag{7.8.V}$$

$$OH^{\cdot} + RH \rightarrow H_2O + R^{\cdot}$$

This is the essence of the mechanism, but some radicals present can dissociate to yield the observed aldehyde and ketone products.

With $RO^{\cdot}$, for example, ketones and alkyl radicals may be formed, e.g.

$$C_3H_7(CH_3)_2CO^{\cdot} \rightarrow C_3H_7COCH_3 + CH_3^{\cdot}$$

$$\rightarrow CH_3COCH_3 + C_3H_7^{\cdot}$$

For $ROO^{\cdot}$, in a typical case, aldehydes can result:

$$CH_3CH_2CH_2OO^{\cdot} \rightarrow CH_3CH_2O^{\cdot} + HCHO^{*} \tag{7.8.VI}$$

$$\rightarrow CH_3O^{\cdot} + CH_3CHO$$

Formaldehyde is unreactive at these temperatures, but acetaldehyde can contribute another branching reaction

$$CH_3CHO + O_2 \rightarrow CH_3CO^{\cdot} + HO_2^{\cdot}$$

However, the main low temperature branching sequence is reactions (7.8.II), (7.8.III) and (7.8.V) and this can be represented:

$$R^{\cdot} + O_2 \rightarrow ROO^{\cdot} \rightarrow ROOH \rightarrow \text{branching} \tag{7.8.1}$$

The effect of increasing temperature is very complicated and only a simplified outline is attempted below.

Region of negative temperature coefficient

To give this change in temperature dependence some process must compete for a

[7] See, for example, M. Seakins and C. N. Hinshelwood, *Proc. R. Soc.* (*London*), **A276**, 324 (1963).

chain carrier in the above branching sequence and so diminish the autocatalytic rate. A likely candidate is:

$$ROO^{\cdot} \rightarrow R^{\cdot} + O_2 \qquad (7.8.VII)$$

which is the reverse of (7.8.II). It increases the stationary concentration of $R^{\cdot}$ and this, with increasing temperature, leads to an alternative reaction of $R^{\cdot}$ with oxygen, for example:

$$CH_3C^{\cdot}HCH_3 + O_2 \rightarrow CH_3CH{=}CH_2 + HO_2^{\cdot}$$

Thus the sequence

$$ROO^{\cdot} \rightleftharpoons R^{\cdot} + O_2 \rightarrow \text{olefin} + HO_2^{\cdot} \qquad (7.8.2)$$

competes with branching sequence (7.8.1), and as temperature increases the overall rate falls.

High temperatures

At some high temperature the relatively unreactive HO_2 becomes involved in hydrogen peroxide formation

$$RH + HO_2^{\cdot} \rightarrow HOOH + R^{\cdot} \qquad (7.8.VIII)$$

and though hydrogen peroxide is more stable than organic peroxides it decomposes at a significant rate above about 700 K:

$$HOOH + M \rightarrow 2OH^{\cdot} + M \qquad (7.8.IX)$$

$$OH^{\cdot} + RH \rightarrow R^{\cdot} + H_2O$$

Hence a new branching sequence:

$$HO_2^{\cdot} \rightarrow HOOH \rightarrow \text{branching} \qquad (7.8.3)$$

supercedes the lower temperature sequences and reaction rates once again rise with temperature.

Reactions at high temperatures, whether explosive or not, lead to increased decomposition of radicals and, overall, the hydrocarbons are degraded to lower molecular weight species giving products similar to hydrocarbon pyrolyses.

Cool flames

This region of the explosion diagram adjoins the ignition boundary in the temperature range that includes the negative temperature coefficients. The phenomenon should involve the low temperature mechanism together with the competing non-branching sequence, (7.8.2).

Reaction starts with the slow branched mechanism and accelerates as ROOH accumulates. It speeds up further as the chain-thermal contribution becomes effective and decomposition of ROOH ensues. The accompanying rise in temperature means the overall reaction may enter the range of negative temperature coefficient with subsequent fall in rate and temperature. The time scale for these events is very short and the result is effectively a sudden rise and fall, or pulse, in temperature and pressure.

Electronically excited formaldehyde is one of the transient products in reactions such as (7.8.VI), and its chemiluminescence gives the visible blue glow of the cool flame.

The rapid pulse so described returns the system almost to the initial conditions, with only a small fraction of reactants consumed. The cycle may be repeated giving further cool flames until consumption of reactants is too high for repetition. A cool flame may, under certain conditions, take the system into the high temperature region where branching sequence (7.8.3) operates. The reaction then continues unchecked to explosion.

A useful insight into these complex systems is provided by considering them as oscillatory reactions showing the kinetic features of this class[8]. Finally, as a dramatic indication of how much more complicated real systems are than the simple outline given here, it is worth noting that oxidation of 3 methyl pentane in the cool flame region gives 86 products[9].

7.9 High Temperature Combustion—Flames

In an explosion the temperature rises to several thousand degrees and conditions are similar to those that prevail in flames, which can be studied in specially constructed burners. Flowing premixed reactants through a burner, which consists of many fine capillaries packed together, gives flat flames. As with other flow systems the reaction time increases along the direction of flow and the variation of temperature and concentrations with 'time from burner' can be measured in several ways. Temperature can be probed with a fine, silica covered thermocouple. Radicals have been detected by e.s.r., absorption spectra, but more usually by emission spectra, and mass spectrometry with a suitable leak into the ionization chamber.

Without entering into detail about the physical structure of flames under stable conditions, three general zones perpendicular to the gas flow may be mentioned—preheat, reaction and recombination zones[10]. In the *preheat zone* considerable degradation of hydrocarbon to lower alkanes, olefins and hydrogen takes place, so in the reaction zone similar results are found with many different hydrocarbon fuels.

In the *reaction zone* ($\sim$0·1 mm thick) the predominant reactions are governed by the effect of high temperature, typically 2000–3000 K. Some major features are:

(i) Rate constants for high activation energy radical–molecule reactions tend to their upper limit value of A_{exp}, the pre-exponential factor, since $e^{-E/RT}$ tends to unity.

(ii) Radical concentrations are very high, approaching 50% of fuel content. Radical–radical reactions have very low activation energies and their rates do not change greatly but with such high concentrations these reactions can dominate the system.

[8] The kinetics of oscillatory reactions are reviewed by B. F. Gray, *Special Periodical Report, Reaction Kinetics*, Vol. 1, Chemical Society, London, 1975, p. 309.
[9] P. Burat, C. F. Cullis and R. T. Pollard, *13th Int. Symp. on Combustion*, The Combustion Institute, 1971.
[10] For comprehensive information see A. G. Gaydon and H. G. Wolfhard, *Flames, Their Structure Radiation and Temperature*, 1970; C. P. Fenimore, *Chemistry in Premixed Flames*, Pergamon, 1964.

(iii) Species such as peroxides and aldehydes that are involved in degenerate branching are pyrolysed and replaced by simple atoms and radicals. Hydrogen atoms become particularly important due to such reactions as:

$$RCHO \rightarrow RCO^{\cdot} + H^{\cdot}$$

$$RCH_2CH_2^{\cdot} \rightarrow RCH{=}CH_2 + H^{\cdot}$$

(iv) In hydrocarbon flames, the chain branching sequence resembles that in the hydrogen–oxygen reaction:

$$H^{\cdot} + O_2 \rightarrow OH^{\cdot} + O^{\cdot\cdot}$$

$$O^{\cdot\cdot} + H_2 \rightarrow OH^{\cdot} + H^{\cdot}$$

$$O^{\cdot\cdot} + RH \rightarrow R^{\cdot} + OH^{\cdot}$$

$$OH^{\cdot} + H_2 \rightarrow H_2O + H^{\cdot}$$

$$OH + RH \rightarrow H_2O + R^{\cdot}$$

By the end of the reaction zone quasi-equilibrium is set up due to the speed of reverse reactions such as

$$O^{\cdot\cdot} + OH^{\cdot} \rightarrow O_2 + H^{\cdot}$$

The *recombination zone* is an extended region with such dominant reactions as:

$$H^{\cdot} + H^{\cdot} + M \rightarrow H_2 + M$$

$$H^{\cdot} + OH^{\cdot} + M \rightarrow H_2O + M$$

and the final composition of the burnt gas is largely governed by the overall equilibrium:

$$H_2 + CO_2 \rightleftharpoons H_2O + CO$$

Non-equilibrium distributions

Many reaction products at these high temperatures are in vibrationally or electronically excited states which decay to their ground state by chemiluminescence. Ionization also exceeds equilibrium values at flame temperatures. Possible sources of ions are:

$$O^{\cdot\cdot} + CH^{*} \rightarrow HCO^{+} + e^{-}$$

$$H_2O + HCO^{+} \rightarrow H_3O^{+} + CO$$

and H_3O^+ is the predominant ion.

Polymeric $C_nH_n{}^+$ ions are also found, perhaps due to:

$$CH^{*} + C_2H_2 \rightarrow C_3H_3{}^{+} + e^{-}$$

and such ions are likely precursors of soot.

Incomplete combustion—soot formation

Although oxidation of organic compounds leads eventually to carbon monoxide and carbon dioxide, fuel-rich flames produce solid carbon and the chemiluminescence of excited carbon gives the intense yellow colour to such flames. Aggregation of carbon results in soot deposits and the carbon formation can be interpreted through the overall equilibrium:

$$2CO \rightleftharpoons CO_2 + C_{solid}$$

This lies to the right at low temperature and to the left above about 1000 K though at low temperature the approach to equilibrium is very slow. The detailed mechanism for production of carbon is not fully understood but in essence hydrocarbons are pyrolysed to low molecular weight fragments such as acetylene and ionization, polymerization and dehydrogenation occur.

7.10 The Internal Combustion Engine

The reaction between a pure hydrocarbon and oxygen is immensely complicated and this is compounded when commercial fuel is burnt in an internal combustion engine. The fuel is a mixture of alkanes, olefins and aromatics mainly in the C_4–C_9 range, with certain additives such as lead tetra-alkyls. The oxidant is air containing nitrogen, of course, as well as oxygen. The 'reaction vessel' is the cylinder with piston, valves and igniting device (spark plugs)[11].

In such a system the reactions taking place are analogous to those considered previously, but combustion is by no means complete. After ignition and combustion in the engine cylinder the partly combusted mixture is suddenly pushed out through the exhaust. Temperature and pressure are drastically and almost instantaneously reduced and concentrations are 'frozen' at their high temperature values, in many cases far above low temperature equilibrium concentrations.

Emitted products include carbon dioxide, water, hydrogen, hydrogen peroxide, peroxyacids, alcohols aldehydes and ketones. Also present, and characteristic of incomplete combustion, are carbon monoxide, polyacetylenes and soot together with some unburnt fuel. The nitrogen in air also reacts giving oxides, especially nitric oxide, and the addition of anti-knock agents results in the emission of lead compounds. Some investigations of anti-knock action by fast reaction techniques are described in section 4.1.

Atmospheric pollution

Lead compounds, oxidized hydrocarbons, unburnt hydrocarbons, soot, carbon monoxide and nitrogen oxides are highly undesirable additions to the air we breathe. Unfortunately chemical and engineering measures that can be adopted to reduce pollutant emission are usually very expensive, give lower engine efficiency and often produce alternative forms of pollution. Catalytic converters added to engine exhaust systems to promote more complete reaction seem the most promising remedies, but recently they have been shown to catalyse the conversion of the very small proportion of sulphur compounds in fuel to sulphuric acid, with deleterious consequences, and removing these traces of sulphur from fuel would be extremely expensive.

We are far from solving the chemical and engineering problems, and even farther from understanding the associated social and economic ramifications.

[11] For a fuller account of combustion systems, see J. N. Bradley, *Flame and Combustion Phenomena*, Science Paperbacks, Chapman and Hall, London, 1972.

7.11 Photochemical Smog

Perhaps the extreme form of atmospheric pollution resulting from the internal combustion engine is the phenomenon of photochemical smog: a visibility-reducing haze with acid smell and taste, that is very irritating to the eyes and has a decidedly unpleasant effect on the lungs. Photochemical smog is found where the following conditions coincide: a high concentration of the primary pollutants (hydrocarbons and oxides of nitrogen); strong sunlight; and a relatively stable air mass resulting from local geography. The greater Los Angeles basin exhibits a classic conjunction of these factors. To have any hope of alleviating such problems the relevant chemistry and kinetics must be understood but again we are confronted with a system of extreme complexity and at present only the main mechanistic features can be outlined[12].

Some of the main features of a typical day's smog are shown in figure 7.6, though some important components such as aerosols are excluded. Research has shown that the major stages in the developing smog are probably:

(i) Emission of hydrocarbons and nitrogen oxides from morning traffic, which reaches a peak during 'rush hour'.

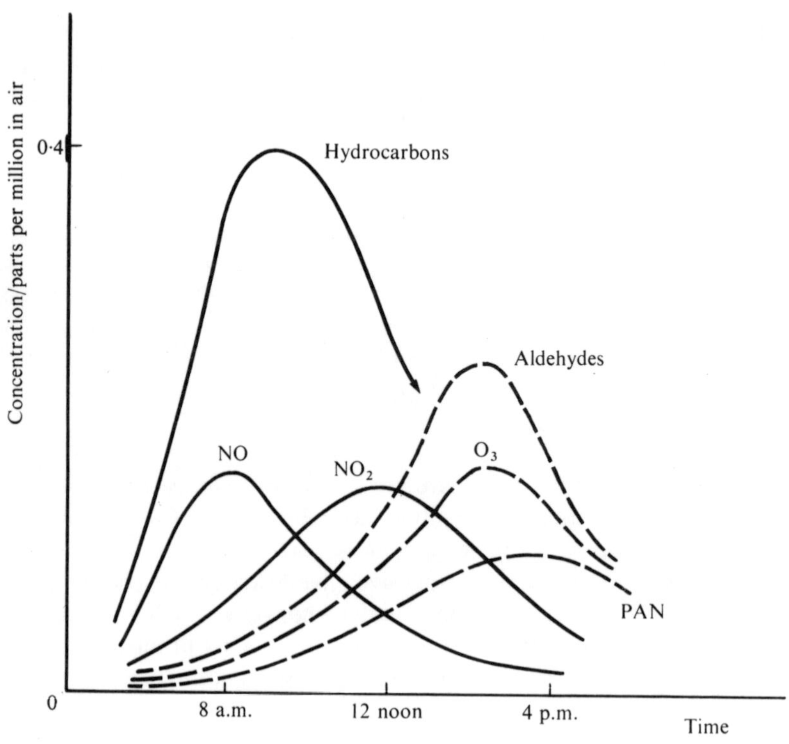

Figure 7.6
The variation of some pollutant concentrations with time throughout a typical day of photochemical smog.

[12] See, for example, R. S. Berry and P. A. Lehman, *Ann. Rev. Phys. Chem.*, **22**, 47 (1971).

(ii) Absorption of sunlight by nitrogen dioxide, which increases as the concentration of oxide and the sun's intensity increase.

(iii) A series of reactions in which nitric oxide is converted to nitrogen dioxide, with simultaneous reactions yielding oxidants such as ozone and aldehydes.

(iv) Oxidation of hydrocarbons to products including eye irritants like PAN (peroxyacyl nitrates) and haze-forming aerosols.

(v) The end of oxidation reactions and dispersal of products.

The stages overlap considerably and each represents the overall reults of many reactions which are not fully understood. Some of the more important are given below.

The main light absorption is by NO_2:

$$NO_2 + h\nu \rightarrow NO + O^{\cdot\cdot}$$

The oxygen atoms react to give ozone:

$$O^{\cdot\cdot} + O_2 + M \rightarrow O_3 + M$$

which can also be removed by:

$$NO + O_3 \rightarrow NO_2 + O_2$$

These reactions cannot account for the known ozone concentration. Oxygen atoms must initiate radical chains in which ozone and other oxidants are formed and nitric oxide converted to nitrogen dioxide. Research indicates that aromatic and olefin hydrocarbons are vital to chain initiation which may occur through, for example:

$$O^{\cdot\cdot} + RCH{=}CHR' \rightarrow RCO^{\cdot} + R'CH_2^{\cdot}$$

The chain then proceeds as in table 7.1, where all alkyl radicals are represented as $R^{\cdot}$ for simplicity. So the chains produce ozone, oxidant radicals and nitrogen dioxide, which can be photolysed again. The 'oxygen atom mechanism' does not suffice to explain the overall rates observed and additional initiation steps involving electronically excited oxygen molecules and hydrogen atoms have been invoked.

Table 7.1
Some chain reactions producing ozone and converting NO to NO_2 in photochemical smog systems.

Ozone producing	Converting NO to NO_2
$R^{\cdot} + O_2 \rightarrow ROO^{\cdot}$	
$ROO^{\cdot} + O_2 \rightarrow RO^{\cdot} + O_3$	$ROO^{\cdot} + NO \rightarrow RO^{\cdot} + NO_2$
	$RO^{\cdot} + O_2 + NO \rightarrow ROO^{\cdot} + NO_2$
$RCO^{\cdot} + O_2 \rightarrow RCO_3^{\cdot}$	
$RCO_3^{\cdot} + O_2 \rightarrow RCO_2^{\cdot} + O_3$	$RCO_3^{\cdot} + NO \rightarrow RCO_2^{\cdot} + NO_2$
$RCO_2^{\cdot} + O_2 \rightarrow RCO^{\cdot} + O_3$	$RCO_2^{\cdot} + NO \rightarrow RCO^{\cdot} + NO_2$

Aldehydes can be formed by decomposition of unstable oxidant radicals, or by interaction between olefins and ozone. PAN (peroxyacyl nitrates) results from the reaction:

$$\underset{\substack{\| \\ O}}{RC}OO^{\cdot} + NO_2 \rightarrow \underset{\substack{\| \\ O}}{RC}OONO_2$$

The strongest eye irritants are where $R^{\cdot}$ is methyl or phenyl.

The haze formed in the smog is due to aerosols, polymers of particle size which scatter light and so reduce visibility. The aerosols are usually formed by a free radical polymerization of straight-chain or cyclic olefin molecules.

As stated above, this section only indicates some of the important reactions occurring in photochemical smog. Solving such environmental problems depends on combining basic and applied research with social and political action. This is not impossible, as shown by the elimination of London smogs in the last twenty years. However, that smog was caused mainly by SO_2 and soot from the burning of solid fuels whereas photochemical smog involves more complicated chemistry and the economics of the motor industry, which certainly present formidable obstacles to progress.

Further Reading

N. N. Semenov, *Some Problems in Chemical Kinetics and Reactivity*, Vol. 2, Pergamon, 1958.

B. Lewis and G. von Elbe, *Combustion Flames and Explosions of Gases*, Academic Press, 1961.

G. J. Minkoff and C. F. H. Tipper, *Chemistry of Combustion Reactions*, Butterworths, 1962.

F. S. Dainton, *Chain Reactions*, Methuen, 1966.

P. Gray and P. R. Lee in *Oxidation and Combustion Reviews*, Vol. 2, (Ed. C. F. H. Tipper), Elsevier, 1967.

R. R. Baldwin and R. Walker in *Essays in Chemistry*, Vol. 3, (Eds J. N. Bradley *et al.*), Academic Press, 1972 (hydrogen–oxygen reaction).

J. N. Bradley, *Flame and Combustion Phenomena*, Science Paperbacks, Chapman and Hall, 1972.

P. A. Leighton, *Photochemistry of Air Pollution*, Academic Press, 1961.

Air Pollution, (Ed. A. C. Stern), McGraw-Hill, 1968.

R. S. Berry and P. A. Lehman, *Ann. Rev. Phys. Chem.*, **22**, 47 (1971).

Exercises

7.1
A reaction undergoes a *thermal* explosion characterized by the equations in section 7.1. Show that for a vessel at temperature, T_s, the gas temperature at the explosion limit, T_c, is given by $T_c - T_s = RT_c^2/E$ where E is the activation energy for the overall reaction.

A unimolecular reaction with pre-exponential factor 10^{16} s^{-1} and activation energy 250 kJ mol^{-1} is 150 kJ mol^{-1} exothermic. The reaction is investigated in a reaction vessel with a surface to volume ratio of 1 cm^{-1} and its coefficient for heat transfer, K, is 4×10^{-4} J cm^{-2} K^{-1} s^{-1}. The vessel is placed in a thermostatted furnace at 700 K. Calculate the critical gas temperature at the explosion limit and the corresponding reactant pressure.

(Ans: $T_c = 717$ K; reactant pressure = 332 Torr)

7.2
In a certain chain reaction the average time between propagation steps is $\bar{t}$. The radical concentration is $[\bar{x}]$ and so the rate of reaction may be expressed: $d[\text{prod.}]/dt = [\bar{x}]/\bar{t}$. The variation of $[\bar{x}]$ with time for a chain reaction with linear branching and termination steps is given by equation (7.2.4) where the branching factor is positive. Show that where $e^{\phi t} \gg 1$

$$\ln[\text{prod.}] = \ln(R_i / \phi^2 E) + \phi t$$

7.3
In a certain branched-chain reaction the pressure changes as the reaction proceeds and ΔP is proportional to the amount of product formed. A specially designed sensitive manometer could record the change in pressure over time intervals as short as a hundredth of a second. In a run at 800 K the pressure change, initially too small to detect, varied with time as shown in the table:

Time/s	0·15	0·16	0·17	0·18	0·19
ΔP/Torr	0·13	0·36	0·97	2·65	7·53

Using the relationship deduced in exercise 7.2 evaluate the branching factor under these conditions.

(Ans: $\phi = 100$)

7.4
For a stoichiometric mixture of hydrogen and oxygen the pressure at the second explosion limit varied with temperature as shown in the table:

P_{II}/Torr	28	40	57	83	115
T/°C	470	490	510	530	550

The mechanism for the reaction is given in section 7.4. If the activation energy of the termination reaction is $-6{\cdot}7$ kJ mol^{-1} what activation energy of the branching reaction is indicated by the data?

(Ans: 82 kJ mol^{-1})

7.5
For the hydrogen–oxygen reaction, $[M]$ in equation (7.4.4) should represent the sum of terms for each species present acting as third body, each term being weighted by a coefficient expressing the relative efficiency of that body.

The second explosion limit varies with the proportions of reactants in the mixture. At 820 K this variation is shown in the table:

Total pressure at limit/Torr	147	135	124	112
$[O_2]:[H_2]$ ratio	6·35	2·38	1·06	0·388

What are the relative efficiencies of hydrogen and oxygen as third body?

(Ans: $H_2:O_2 = 1{\cdot}57:1$)

7.6
A simplified mechanism for carbon monoxide combustion is

1	$CO + O_2$	$\rightarrow CO_2 + O^{\cdot\cdot}$
2	$O^{\cdot\cdot} + CO$	$\rightarrow CO_2{}^*$
3	$CO_2{}^* + O_2$	$\rightarrow CO_2 + 2O^{\cdot\cdot}$
4	$CO_2{}^*$	$\rightarrow$ wall
5	$CO_2{}^* + M$	$\rightarrow CO_2 + M$

Write down an expression for the branching factor and show that at the first limit $k_3[O_2] = k_4$ and at the second $k_3[O_2] = k_4[M]$.

7.7
The oxidation of phosphorus vapour in the presence of some solid phosphorus occurs by a branched-chain mechanism and there are two explosion limits. For the upper limit $P_{II} = (\text{constant})\ e^{-E/RT}$ where $E = 42$ kJ mol^{-1}. The overall reaction in the gas phase is:

$$P_4 + 5O_2 \rightarrow P_4O_{10}$$

and a suggested mechanism is, with radical species underlined:

1	$P_4 + O_2$	$\rightarrow \underline{P_4O^{\cdot}} + \underline{O^{\cdot\cdot}}$
2	$P_4 + \underline{O^{\cdot\cdot}} + M$	$\rightarrow \underline{P_4O^{\cdot}} + M$
3(a)	$\underline{P_4O^{\cdot}} + O_2$	$\rightarrow \underline{P_4O_2{}^{\cdot}} + \underline{O^{\cdot\cdot}}$
3(b)	$\underline{P_4O_2{}^{\cdot}} + O_2$	$\rightarrow \underline{P_4O_3{}^{\cdot}} + \underline{O^{\cdot\cdot}}$
		$\vdots$
3(h)	$\underline{P_4O_8{}^{\cdot}} + O_2$	$\rightarrow \underline{P_4O_9{}^{\cdot}} + \underline{O^{\cdot\cdot}}$
3(i)	$\underline{P_4O_9{}^{\cdot}} + O_2$	$\rightarrow P_4O_{10} + \underline{O^{\cdot\cdot}}$
4	$\underline{O^{\cdot\cdot}} + O_2 + M$	$\rightarrow O_3 + M$
5	$\underline{O^{\cdot\cdot}}$	$\rightarrow$ wall

Using the steady-state hypothesis in the manner described for the hydrogen–oxygen reaction in section 7.4 (leading up to equation (7.4.7)) show that

$$8k_2[P_4][M] - k_4[O_2][M] - k_5 = 0$$

The vapour pressure of phosphorus is proportional to $e^{-\lambda/RT}$ where λ is the heat of vapourization. Given that the activation energies for reactions 2 and 4 are zero and -8 kJ mol^{-1}, evaluate λ from the conditions at the upper limit.

(Ans: $\lambda = 34$ kJ mol^{-1})

Chapter 8: Molecular Dynamics

Kinetic systems discussed in previous chapters have ranged from mixed hydrocarbons burning in air to elementary radical reactions. The speed of such reactions has been described by their rate constant, but a moment's thought shows that this is not the most fundamental of kinetic parameters since it conceals immense complexity at the molecular level. It represents the net result of perhaps 10^{26} bimolecular collisions per cm^3 per second in the gas phase. The nature and result of the individual collisions vary widely but the rate constant is still a good description of bulk molecular behaviour at a given temperature because such an enormous number of collisions ensures very well-defined averages. For example, the relative velocity of two colliding molecules may have almost any value but the velocities in the bulk gas conform to the Maxwell–Boltzmann velocity distribution and average relative velocity on collision is defined accurately.

With the advent of new experimental methods and complementary advances in theory, kinetics has been extended to the new world of single collisions and reactions can be studied in hitherto undreamt of molecular detail. It is now feasible for the experimenter to investigate collisions between reactants in specific internal energy states interacting at a selected relative velocity, which means a selected translational energy—even the relative orientation of the colliding species may sometimes be controlled. Such experiments measure the likelihood of either energy transfer or reaction on collision, but in this text we concentrate on reactive collisions. From the angle and velocity with which reaction products emerge from collisions it is possible to deduce whether the products are formed in direct, very short-lived interactions or from intermediate collision complexes, and to calculate the distribution of any reaction exothermicity between translational and internal excitation of products. Finally the familiar bulk kinetic properties can be calculated from these results for single collisions at defined energy.

Reaction theory has also advanced greatly in recent years, especially with computer simulation of reactive collisions. Taking a suitable potential energy surface for the system the reaction probability can be computed for a given set of initial conditions. Then appropriate averaging procedures yield theoretical predictions which can be compared directly with the results of the detailed kinetic experiments.

It should be evident from this brief introduction that the kinetic parameters for bulk reactions—rate constants and activation energies—are just not applicable to single collisions at defined energies. However, appropriate new parameters arise naturally when one takes a detailed look at the most important experimental technique in this field, crossed molecular beams, which are described in experimental detail in section 9.1.

8.1 New Kinetic Parameters: Reaction Cross Section

Consider a beam of reactant A with concentration (number of species per unit volume) n_A and let the beam have unit area perpendicular to the beam direction. This beam then passes through a region containing reactant B with concentration n_B—for example in another beam crossing the first at right angles. Let all A have relative velocity v with respect to B. The intensity of the A beam is $I_A = n_A v$. In passing through the crossing region where collision with B can take place the intensity in the direction of the A beam is reduced since some A molecules are scattered out of the beam due to such collisions. The beam concentrations are chosen low enough to ensure that a reactant molecule will collide only once in the crossing region, so scattering results refer to *single* collisions.

Consider the loss in intensity, $-\mathrm{d}I_A$, due to collisions in the length $\mathrm{d}l$ along the crossing region whose total length is L. These quantities are illustrated in figure 8.1. The loss of intensity must be proportional to the incident intensity $I_A(l)$ at this layer, to the thickness $\mathrm{d}l$ and to the concentration of B. Thus:

$$-\mathrm{d}I_A = \sigma(v)\, I_A(l)\, n_B\, \mathrm{d}l \qquad (8.1.1)$$

The proportionality constant $\sigma(v)$ is a measure of the probability that an A molecule is scattered from the beam of unit area after a single collision with B. This scattering may result from elastic (translational–translational) or inelastic (translational–internal) energy transfer or from reactive collisions and so $\sigma(v)$ refers to the *total scattering probability*. The total probability will vary with the relative velocity at which the species collide.

The net loss of A in passing through the entire width of the crossing region is the sum of all scattering from $l = 0$ to $l = L$. Thus, if the incident intensity is $I_A{}^0$ at $l = 0$ and the final intensity $I_A{}^{\mathrm{f}}$ at $l = L$

$$-\int_{I_A{}^0}^{I_A{}^{\mathrm{f}}} \frac{\mathrm{d}I_A}{I_A} = \int_0^L \sigma(v) n_B\, \mathrm{d}l$$

$$\ln \frac{I_A{}^0}{I_A{}^{\mathrm{f}}} = \sigma(v) n_B L \qquad (8.1.2)$$

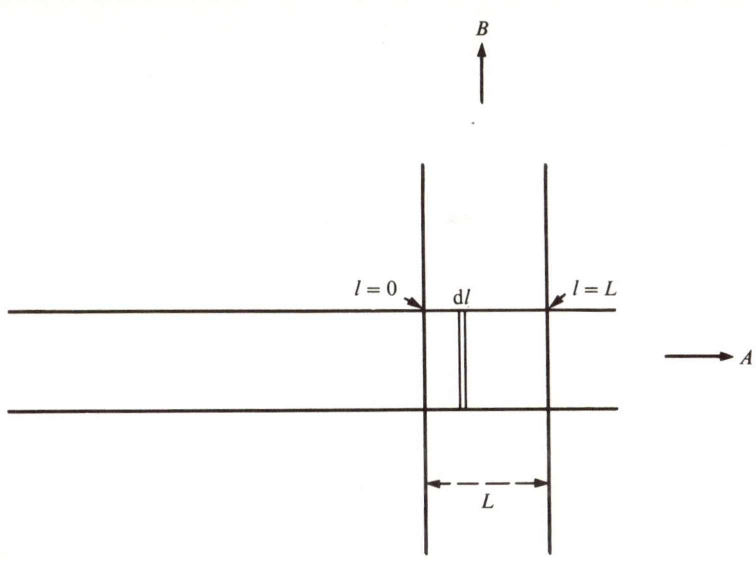

Figure 8.1
Diagram of a beam of A molecules, crossing a beam of B molecules, width L. The intensity of the A beam is $I_A{}^0$ at $l = 0$, $I_A(l)$ at element dl, and $I_A{}^f$ at $l = L$.

Equation (8.1.2) shows a formal similarity to the Beer–Lambert law* which describes the loss of light intensity on passing through a light-absorbing system. Inspection of equation (8.1.2) shows that the proportionality constant which expresses the total scattering probability has the dimensions of area—it is the 'effective' cross sectional area for the collisional process and is known simply as the *collisional cross section* (for total scattering). The collisional cross section can be considered as the sum of cross sections for elastic scattering, $\sigma_{EL}(v)$, inelastic scattering, $\sigma_{IN}(v)$, and reactive scattering, $S(v)$, and experimentally it is possible to measure these cross sections separately. Cross sections such as the *reaction cross section* $S(v)$ which vary with relative velocity can also be expressed as a function of relative translational energy, ε, of the colliding molecule—$S(\varepsilon)$.

8.2 Reaction Cross Section and Reaction Rate at Single Velocity

It may be useful at this stage to indicate the relationship between the new kinetic parameter, reaction cross section, and the familiar property of the bulk gas system, the rate constant. The frequency of reactive collisions, all at single relative velocity v, can be calculated by the device employed for the Simple Collision Theory (see section 5.1, bearing in mind that the velocity considered there is the *average* relative velocity). Let one A molecule move with relative velocity v, B molecules being considered stationary. Figure 8.2 shows the imaginary cylinder of length v and cross sectional area

* See equation (2.4.1).

equal to $S(v)$, the reaction cross section at this relative velocity. In one second the A molecule collides with all B molecules in the cylinder. The collision frequency for reaction is thus the volume of the cylinder multiplied by the concentration of B

$$Z_B(v) = S(v)\, v\, n_B$$

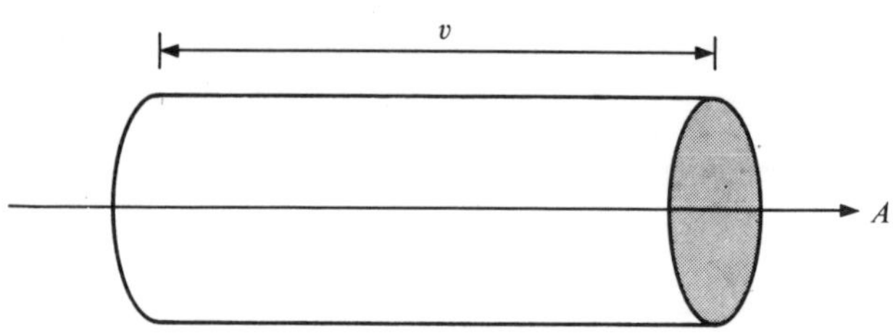

Figure 8.2
Showing the path of a molecule of A moving with relative velocity v, B molecules being stationary. The cylinder has length v and its cross sectional area (shaded) is $S(v)$ the reaction cross section at relative velocity v.

The total frequency of reactive collisions for all A molecules, $Z_{AB}(v)$, is $Z_B(v)$ times the concentration of A, and this is also the rate of reaction. Hence:

$$-\mathrm{d}n_A/\mathrm{d}t = -\mathrm{d}n_B/\mathrm{d}t = Z_{AB}(v) = S(v)\, v\, n_A n_B \tag{8.2.1}$$

The reaction rate for bimolecular collisions at this relative velocity can be expressed in terms of a rate constant *for a single relative velocity*, $k(v)$

$$-\mathrm{d}n_A/\mathrm{d}t = k(v) n_A n_B \tag{8.2.2}$$

and comparison of (8.2.1) and (8.2.2) shows:

$$k(v) = S(v)v \tag{8.2.3}$$

The next stage in examining the nature of cross sections is to consider how the collisional cross section (for total scattering) and the reaction cross section vary with velocity, or translational energy[1].

8.3 Variation of Collisional Cross Section with Translational Energy

The interaction between two colliding molecules is governed by their relative velocity and how the potential energy of the two species varies with their distance apart. A molecule will not be deflected if it passes another at a distance for which the potential energy of interaction is negligible. For distances apart where the potential energy is significant, the amount of deflection will vary with the relative velocity (and hence translational energy) of the colliding species. It is such factors that govern the variation of the kinetic parameter, the cross section, with energy.

[1] A more detailed account along these lines is given by E. F. Greene and A. Kupperman, *J. Chem. Ed.*, **45**, 361 (1968).

This variation must be a vital factor governing molecular behaviour and some relevant general features can be deduced from the way potential energy changes with internuclear distance in the simple case of monatomic reactants (or molecules considered as spheres). This energy is conveniently described by the potential energy function $V(r)$ where r is the internuclear distance. Figure 8.3(a) shows $V(r)$ for simple hard spheres, whilst figures 8.3(b) and (c) represent more realistic potentials discussed later.

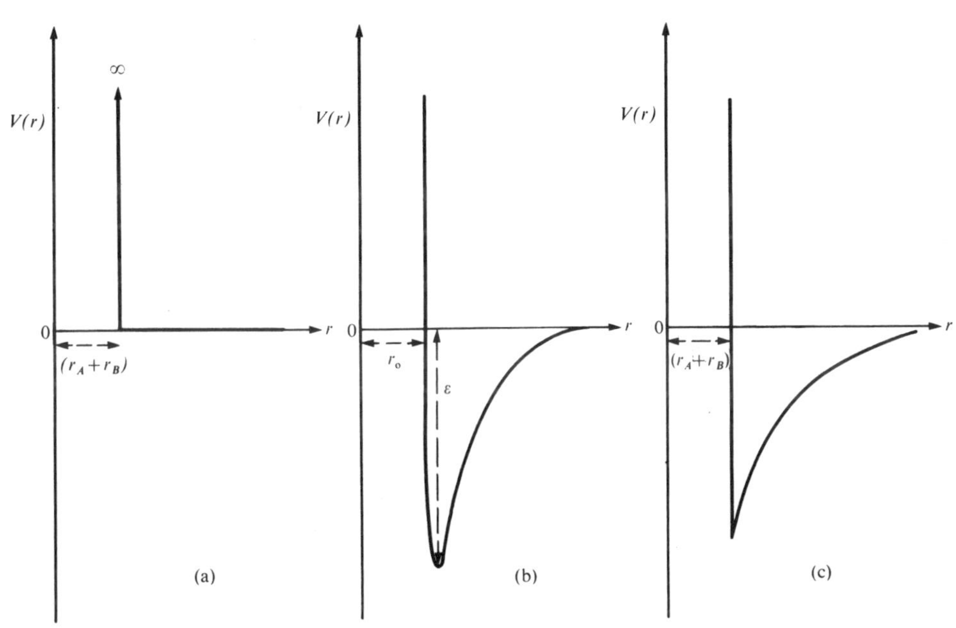

Figure 8.3
Variation of potential energy with internuclear distance: (a) simple hard spheres of radii r_A, r_B; (b) Lennard-Jones type; (c) hard spheres with extra attractive term.

Simple hard spheres

Here $V(r) = \infty$ between $r = 0$ and $r = (r_A + r_B)$ where r_A, r_B are the hard-sphere radii, and $V(r) = 0$ for $r > (r_A + r_B)$.

Figure 8.4(a) shows a collision between A and B in more detail. B is stationary and A travels with the relative velocity v in the direction shown. A line has been drawn through the centre of B parallel to the relative velocity vector. The displacement of A from the line through B is the *impact parameter*, b, a very useful function in describing molecular collisions. As a result of the collision A is scattered through angle θ.

In the simple case where scattering does not depend on any relative orientation of A and B, there is cylindrical symmetry about the line through the centre of B, i.e. $b = 0$. The diagram could be rotated about this line as axis. Thus the contribution,

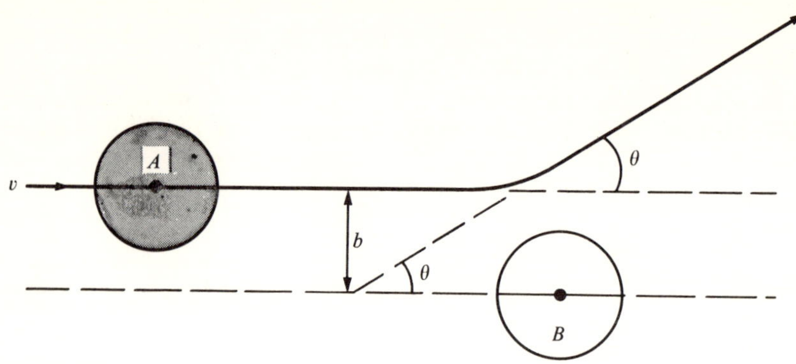

Figure 8.4
(a) Collision between A and B at relative velocity v. b is the impact parameter.

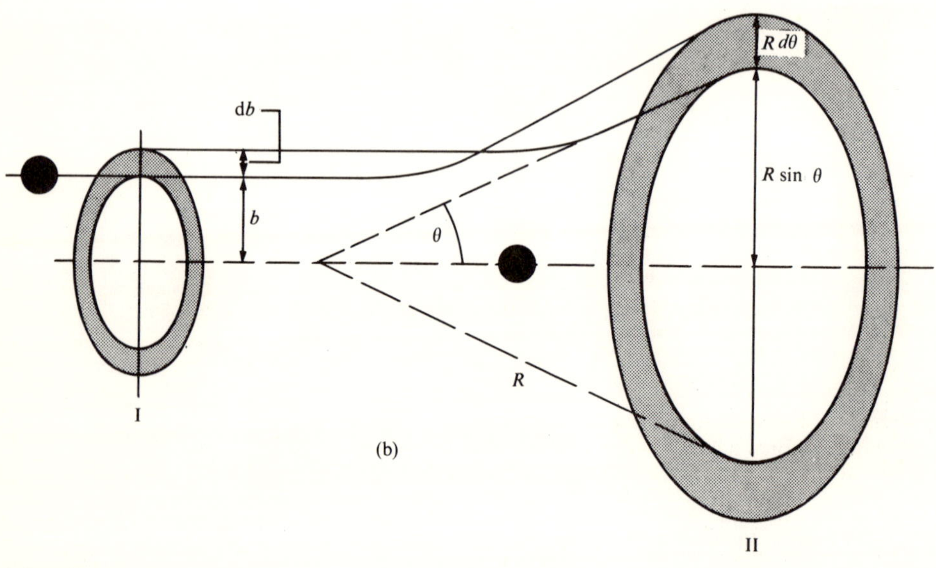

(b) Geometry of cylindrically symmetrical scattering. For clarity, A and B are represented as small black spheres.
The contribution of impact parameters from b to $b + \mathrm{d}b$ is the area of disc I, $2\pi b\,\mathrm{d}b$.
The cone of solid angle between θ and $\theta + \mathrm{d}\theta$ is $\mathrm{d}\Omega$.
By definition, $\mathrm{d}\Omega =$ (area of disc II) / $R^2 = 2\pi\,(R\sin\theta)(R\,\mathrm{d}\theta)\,/\,R^2$.
Therefore, $\mathrm{d}\Omega = 2\pi\sin\theta\,\mathrm{d}\theta$.
The elements $\mathrm{d}b$ and $\mathrm{d}\theta$ are exaggerated for clarity.

$\mathrm{d}\sigma$, of impact parameters from b to $(b + \mathrm{d}b)$ to the collisional cross section is the area of ring-shaped disc (I) in figure 8.4(b), i.e. $2\pi b\,\mathrm{d}b$:

$$\mathrm{d}\sigma = 2\pi b\,\mathrm{d}b$$

Cross sections have hitherto been expressed as functions of relative velocity, but they are easily transformed to functions of molecular translational energy, $\sigma(\varepsilon)$, since

$$\varepsilon = \tfrac{1}{2}\mu v^2$$

When only the initial velocities, or translational energy, of the colliding particles are specified, collisions will occur with all possible values of b, from the minimum to maximum that yield collision. Hence the collisional cross section is

$$\sigma(\varepsilon) = \int_{b_{\text{min}}}^{b_{\text{max}}} 2\pi b \, \mathrm{d}b \tag{8.3.1}$$

For simple hard spheres $b_{\text{min}} = 0$ and $b_{\text{max}} = (r_A + r_B)$, because for $b > (r_A + r_B)$ there is no interaction. Substitution into (8.3.1) gives:

$$\sigma(\varepsilon) = \pi b_{\text{max}}^2 = \pi(r_A + r_B)^2$$

The collisional cross section is thus the simple hard-sphere cross sectional area.

More realistic potentials

Real interactions are governed by repulsive and attractive intermolecular forces, and a good representation is the Lennard-Jones potential:

$$V(r) = 4\varepsilon\left[\left(\frac{r_0}{r}\right)^{12} - \left(\frac{r_0}{r}\right)^{6}\right] \tag{8.3.2}$$

where r_0 and ε are defined in figure 8.3(b).

In equation (8.3.2) the first term is the repulsive potential, the second the attractive potential which is significant over much greater distances—the attractive term is longer range than the repulsive.

The Lennard-Jones potential is rather complicated for the purposes of illustration and a convenient simplification is shown in figure 8.3(c). This potential is:

$$V(r) = \begin{cases} \infty & \text{for } r \text{ between } 0 \text{ and } (r_A + r_B) \\ -\,\text{constant}/r^n & \text{for } r > (r_A + r_B) \end{cases}$$

It is equivalent to hard-sphere scattering from $b = 0$ to $b = (r_A + r_B)$, but there is also some scattering at $b > (r_A + r_B)$ due to the long-range attraction, proportional to $1/r^n$.

This has important consequences for the way collisional cross section varies with energy. With hard spheres, as shown in figure 8.5(a), there is no variation and the cross section is a constant, $\pi(r_A + r_B)^2$. For a more realistic potential, long-range attraction means that the cross section is greater than the hard-sphere value. Furthermore, whereas at very low translational energies the deflection at high impact parameters may be large, as the energy increases the deflection will diminish until eventually it is negligible whenever $b > (r_A + r_B)$, so the cross section tends to the hard-sphere value at high energy. This variation of cross section with translational energy is illustrated in figure 8.5(b).

Differential cross section and collisional cross section

The differential cross section is defined as the cross section for scattering into an

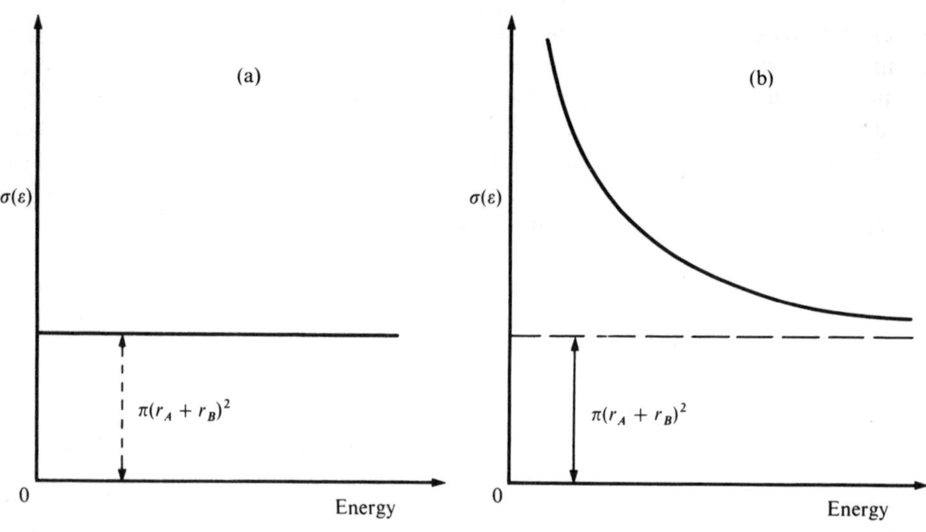

Figure 8.5
Variation of total scattering cross section $\sigma(\varepsilon)$ with translational energy.
(a) simple hard spheres, showing constant cross section, $\pi(r_A + r_B)^2$.
(b) hard spheres with extra attraction term, the cross section falling to the hard-sphere value at high energy.

element of solid angle $d\Omega$ at a specific value of the observed scattering angle θ. Hence the relationship between collisional and differential cross section $\sigma(\varepsilon, \theta)$ is:

$$\sigma(\varepsilon) = \int \sigma(\varepsilon, \theta)\, d\Omega$$

Where scattering does not depend on the orientation of colliding species there is cylindrical symmetry about the line through the centre of B shown in figure 8.4(a) and at angle θ this gives a cone of scattering as shown in figure 8.4(b). The ring-shaped solid angle $d\Omega$ lies between cones at angle θ and at angle $\theta + d\theta$ and elementary geometry shows that $d\Omega = 2\pi \sin\theta\, d\theta$ as illustrated in figure 8.4(b). Hence

$$\sigma(\varepsilon) = \int_0^{\pi} \sigma(\varepsilon, \theta) 2\pi \sin\theta\, d\theta \qquad (8.3.3)$$

It would be extremely useful to establish a relationship between angular distributions and the influence of impact parameter b, since it is impossible, experimentally, to control the value of b and see how scattering depends upon it. All trajectories leading to scattering in the range θ to $\theta + d\theta$ have impact parameters in the range b to $b + db$. Then respective contributions $d\sigma$ to the collisional cross section give:

$$d\sigma = \sigma(\varepsilon, \theta) 2\pi \sin\theta\, d\theta = 2\pi b\, db$$

Hence

$$\sigma(\varepsilon, \theta) = b / \{\sin\theta (d\theta / db)\}$$

8.4 Reaction Cross Section, Reaction Probability

Reaction cross sections can also be considered in more detail as functions of impact parameter. Since only a fraction, usually a small fraction, of collisions results in reaction, it is imperative that the probability of reaction for the collision be taken into account. Let us define the *reaction probability* quite simply as the fraction of collisions with impact parameter b that lead to reaction. This probability may be symbolized as $P(b)$. In line with the optical analogies so popular in scattering studies $P(b)$ is often referred to as the *opacity function*. Since the maximum fraction of collisions that can result in reaction is 1 the value of $P(b)$ must lie between 0 and 1. $P(b)$ would be expected to have a significant value when b is in the range of distances where intermolecular forces are effective, and $P(b)$ should fall to zero above some maximum value of b, b_{max}. The opacity function calculated for an important case is illustrated in figure 8.10.

By analogy with equation (8.3.1) the contribution to reaction cross section of collisions at b may be written as $2\pi bP(b)\,\mathrm{d}b$ and so the reaction cross section as a function of energy alone is

$$S(\varepsilon) = \int_0^{b_{\text{max}}} 2\pi bP(b)\,\mathrm{d}b \qquad (8.4.1)$$

One interesting feature revealed by this expression is that because of the term $2\pi b$ in the area element, contributions to the reaction cross section at higher b values are more heavily weighted. A reaction whose opacity function extends to high b values will have a particularly large cross section.

Detailed reaction cross sections

For experiments which detect the variation of product intensity with scattering angle θ it is useful to define the *differential reaction cross section*, $S(\varepsilon, \theta)$. Then, for the reaction cross section, by analogy with equation (8.3.3)

$$S(\varepsilon) = \int_0^{\pi} S(\varepsilon, \theta) 2\pi \sin\theta\,\mathrm{d}\theta \qquad (8.4.2)$$

Trajectories leading to scattering in the range θ to $\theta + \mathrm{d}\theta$ have impact parameters in the range b to $b + \mathrm{d}b$, and it is possible, from angular distribution studies, to obtain information on how scattering depends on b, and how the opacity $P(b)$ varies with b.

So far cross sections have been defined for a specific relative velocity and so specific translational energy on collision. This implies an equilibrium distribution for the internal energy states of reactants, which is often the case experimentally. However, it is possible to consider collisions between species in single internal energy states, for example in the reaction

$$A(i) + B(j) \xrightarrow{\varepsilon} \text{products}$$

where the reactants are in internal energy states i, j respectively and collide at relative translational energy, ε. The cross section for this reaction may be symbolized $S(i, j, \varepsilon)$. These detailed cross sections are related to the total reaction cross section by summation over the equilibrium internal energy distributions:

$$S(\varepsilon) = \sum_i \sum_j S(i, j, \varepsilon)$$

8.5 Variation of Reaction Cross Section with Energy: Threshold Energy

Collisional cross sections increase as energy decreases, but for reactions some minimum energy is usually necessary. Reaction cross sections should be zero until this minimum energy is reached, whereupon their value increases with energy. However, when energy becomes very high the probability of obtaining the required products must decrease as higher energy reaction modes become more probable. A typical cross section might vary with energy as shown in figure 8.6. The minimum energy for reaction is indicated as ε_0, and this is an experimentally measurable and precisely defined theoretical quantity. It is called the reaction *threshold energy*—the minimum relative translational energy for reaction. The threshold energy is not the same as the experimental activation energy or the barrier height on the potential energy surface for reaction, but it is usually similar in magnitude to these parameters and their relationship is discussed in more detail below. Typical maximum values for reaction cross sections are of the order of combined molecular sizes, up to about 1 nm^2.

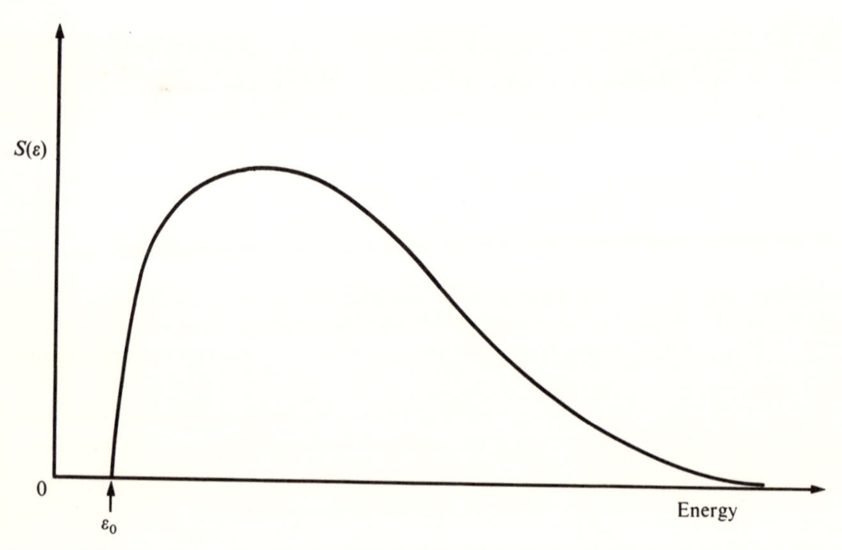

Figure 8.6
The variation of reaction cross section with energy—the excitation function for a reaction. The minimum relative translational energy for reaction is ε_0, the threshold energy.

The curve for reaction cross section versus energy, as shown in figure 8.6, is sometimes called the *excitation function* and it is of truly fundamental importance in chemical kinetics. It provides the basis for discussing reaction probability in single collisions at defined energy, and its relationship to the rate constant and activation energy gives a profound insight into kinetics in conventional bulk gas kinetic systems.

8.6 Reaction Cross Section and Rate Constant

It was shown in section 8.2 that for bimolecular reactions taking place at the same relative velocity, v, the rate constant $k(v)$ is

$$k(v) = S(v)v \tag{8.2.3}$$

In bulk gases molecules collide at all possible relative velocities and there is a term similar to (8.2.3) for each velocity. All these terms contribute to the conventional rate constant ($k(T)$, a function of temperature), each being weighted by the fraction of collisions occurring at that relative velocity. Thus:

$$\begin{aligned} k(T) &= f_1\, k(v_1) + f_2\, k(v_2) + f_3\, k(v_3) + \ldots \\ &= f_1\, S(v_1)v_1 + f_2\, S(v_2)v_2 + f_3\, S(v_3)v_3 + \ldots \end{aligned}$$

where f represents the fraction of collisions occurring at the corresponding velocity. The expression may be written:

$$k(T) = \int_0^\infty f(v, T)\, v\, S(v)\, \mathrm{d}v \tag{8.6.1}$$

The weighting factor for each relative velocity is given by the Maxwell–Boltzmann distribution for relative velocity, symbolized by $f(v, T)$ in equation (8.6.1). This distribution is, of course, a function of temperature. The terms in the integral are illustrated in figure 8.7: figure 8.7(a) shows the distribution function $f(v, T)$; the velocity v; and their product $f(v, T)v$; figure 8.7(b) shows $f(v, T)v$ again; the variation of reaction cross section with velocity, $S(v)$—the excitation function in terms of relative velocity; and the final product, $f(v, T)vS(v)$. This product is the *distribution function for reactive collisions*, and like $S(v)$ its value is zero below the velocity that corresponds to the threshold energy. The area under this curve, shaded in the diagram, is the value of the integral in equation (8.6.1). *So this area is the value of the rate constant.*

These functions can be re-expressed in terms of relative translational energy ($\frac{1}{2}\mu v^2$) and their shapes are similar to those in figure 8.7(b). Only reactive collisions, those with energies above the threshold, contribute to the rate constant, but the reactive collision distribution is fairly narrow so molecules with very high energies contribute little. In general, reactions with low thresholds have higher rate constants than high threshold reactions at a given temperature. Increasing temperature increases the spread of the Maxwellian distribution, and a higher fraction of collisions occur above the threshold. This also enlarges the reactive collision distribution, yielding a higher rate constant value. This result is illustrated for a typical case in figure 8.7(c).

Although cross section measurements allow calculation of the rate constant, it is virtually impossible to reverse this procedure and deduce information about the cross section from rate constants and their temperature variation—it is impossible to 'unfold' the excitation function from the rate constant.

Returning to equation (8.6.1), substituting the standard form of the Maxwellian distribution and transforming from relative velocity to relative translational energy gives the mathematical form of the distribution function for reactive collisions illustrated in figure 8.7(c):

$$k(T) = \left(\frac{1}{\pi\mu}\right)^{1/2} \left(\frac{2}{\bar{k}T}\right)^{3/2} \int_0^\infty S(\varepsilon)\varepsilon\, \mathrm{e}^{-\varepsilon/\bar{k}T}\, \mathrm{d}\varepsilon \tag{8.6.2}$$

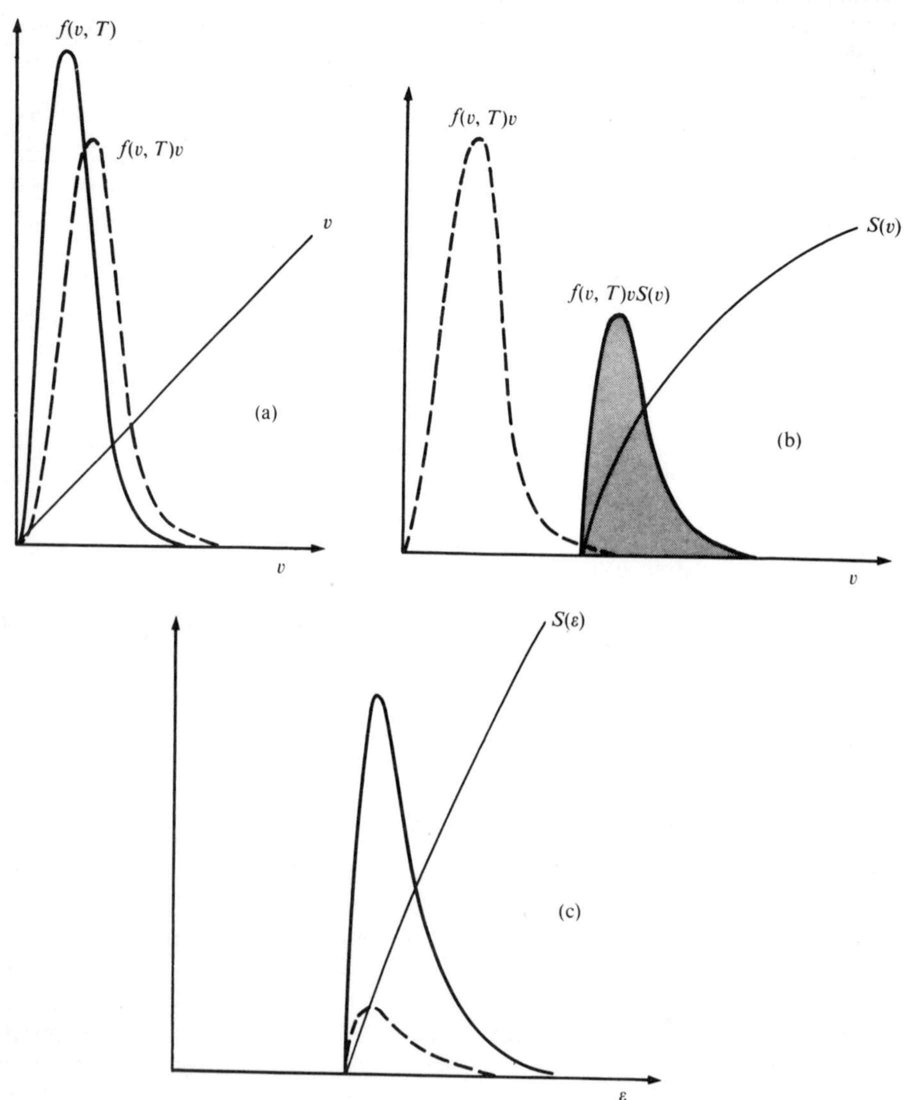

Figure 8.7
(a) The velocity distribution, $f(v, T)$, the relative velocity v and their product $f(v, T)v$.
(b) $f(v, T)v$, an excitation function (reaction cross section $S(v)$ against v) and their product $f(v, T)vS(v)$. The resulting curve is the reactive collision distribution, and the area under the curve (shaded) is the value of the rate constant.
(c) Variation with energy of a reaction cross section $S(\varepsilon)$, and the corresponding distribution function for reactive collisions at two temperatures, 300 K (broken line) and 350 K (solid line).

The diagram demonstrates the essential shapes of these terms and the axes are in *arbitrary* units.

Thus, where $S(\varepsilon)$ is known from theory or experiment, $k(T)$ may be calculated. To give the complete expression for the rate constant of a specific reaction, the dependence of $S(\varepsilon)$ on energy in that case must be expressed by a suitable equation. The resulting relationship will exhibit an appropriate exponential temperature dependence. This leads to the question of the connection between experimental activation energy and threshold energy.

8.7 Experimental Activation Energy and Threshold Energy

The experimental activation energy is conveniently given by:

$$E_{exp} = RT^2\left(\frac{\mathrm{d}[\ln k(T)]}{\mathrm{d}T}\right)$$

From equation (8.6.2) using *molar* energy, E, instead of molecular, ε, and gas constant per mole, R, it can be shown that

$$E_{exp} = \frac{\int_0^\infty S(E)E^2\,\mathrm{e}^{-E/RT}\,\mathrm{d}E}{\int_0^\infty S(E)E\mathrm{e}^{-E/RT}\,\mathrm{d}E} - \frac{3}{2}RT \tag{8.7.1}$$

This relationship was derived by Tolman[2] as long ago as 1920, but it was many years before its significance was widely appreciated. On the right-hand side of (8.7.1), the first term defines the average energy of reactive collisions $\bar{E}_R$, $S(E)$ being zero below the threshold energy, E_0. The second term is simply the average energy of all collisions in the Maxwellian gas, $\bar{E}$. Thus:

$$E_{exp} = \bar{E}_R - \bar{E} \tag{8.7.2}$$

The significance of threshold energy becomes clearer with a graphical representation of equation (8.7.2). Figure 8.8 shows the Maxwellian distribution of relative translational energy, with $\bar{E}$ indicated. The other curve is the distribution of reactive collisions for a typical reaction drawn as a function of energy and with the average energy of reactive collisions shown. There is no general functional relationship between E_0 and E_{exp} since E_0 depends on the precise form of the excitation function, but usually E_0 and E_{exp} should be similar in magnitude, as in the case shown in figure 8.8, where E_{exp} is slightly larger. Both $\bar{E}_R$ and $\bar{E}$ vary with temperature, and so will the activation energy. However, the increase in the average energies should not differ greatly, and variations in activation energy from this cause should be too small to detect experimentally[3].

8.8 Molecular Dynamical Treatment of Reactive Collisions

For the reaction:

$$A + BC \rightarrow AB + C$$

if the forces between atoms are known then, in principle, the chemical reaction becomes a problem in quantum or classical mechanics. The reaction is the change from the situation where r_{AB} and r_{AC} are large and r_{BC} is at the B—C bond length, to the

[2] R. C. Tolman, *J. Am. Chem. Soc.*, **42**, 2506 (1920).
[3] R. L. Wolfgang and A. Menzinger, *Angew Chem.*, **8**, 438 (1969).

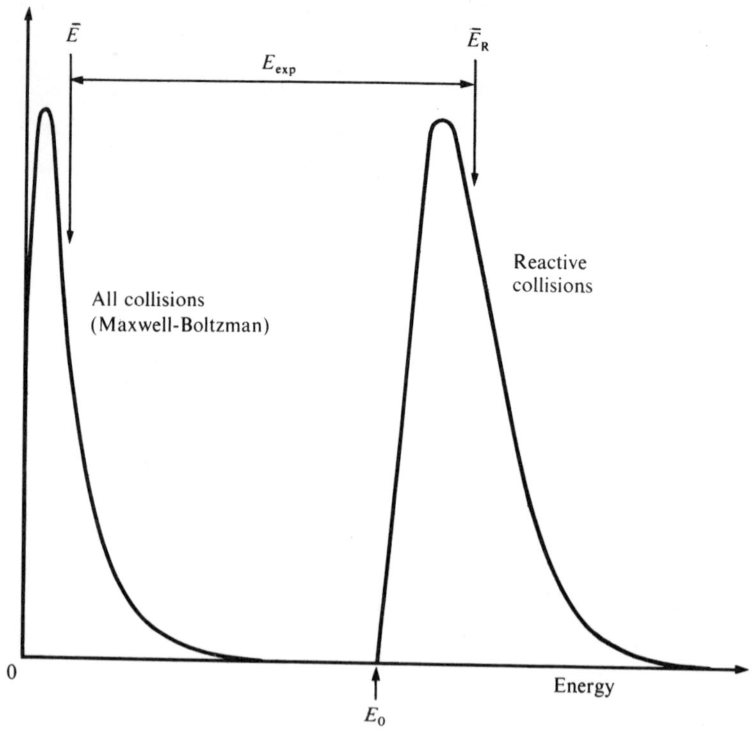

Figure 8.8
The Maxwell–Boltzmann distribution for relative translational energy, and the distribution of reactive collisions expressed as a function of molar energy. The average energy of all collisions, $\bar{E}$, the average energy of reactive collisions, $\bar{E}_R$, the experimental activtaion energy ($E_{exp} = \bar{E}_R - \bar{E}$), and the threshold energy E_0 are shown. The axes are in arbitrary units.

situation where r_{AC} and r_{BC} are large and r_{AB} is the A—B bond length—the reaction may be considered as the variation with time of the internuclear distances r_{AB}, r_{BC} and r_{AC}. The forces can be represented by the potential energy surface which, as shown in section 5.4, is usually calculated from quantum mechanical equations with considerable empirical influence. Fortunately the problem of how nuclei move during collision, their motion 'over' the surface, does not require a quantum mechanical description. Classical mechanics gives a fair degree of accuracy for such molecular trajectories. For a chosen set of initial reactant conditions a computer is used to solve the classical equations of motion so giving the variation of internuclear distance with time[4]. Any one calculation may result either in reaction as portrayed in figure 8.9(a) or in non-reaction, figure 8.9(b). When the 'theoretical experiment' is repeated many times, over a suitable range of initial conditions, the reaction probability at the chosen reactant energy can be calculated and expressed as the reaction cross section.

[4] The history of such calculations is reviewed in K. J. Laidler, *Theories of Chemical Reaction Rates*, McGraw-Hill, 1969.

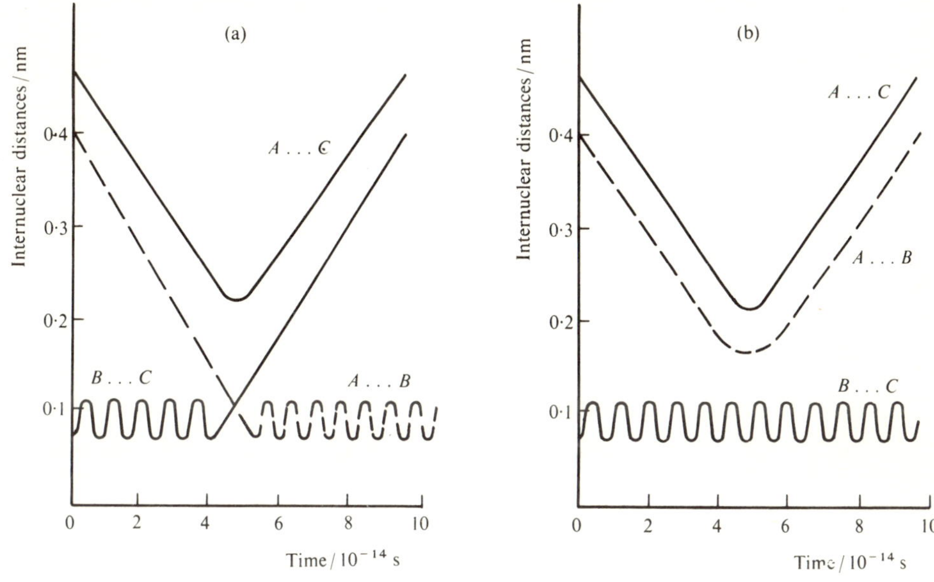

Figure 8.9
Schematic representation of the variation of internuclear distance with time in collisions of A with BC.
(a) Reactive collision $A + BC \rightarrow AB + C$;
Initially, as A approaches BC, the $A \ldots B$ and $A \ldots C$ distances decrease while the $B \ldots C$ distance oscillates due to the vibration of the BC molecule. After reactive collision atom C leaves the product molecule AB, the $A \ldots C$ and $B \ldots C$ distances increase and $A \ldots B$ oscillates with the vibration of the product molecule.
(b) Non-reactive collision $A + BC \rightarrow A + BC$;
The $A \ldots B$ and $A \ldots C$ distances decrease and then increase after collision, and the $B \ldots C$ distance shows the oscillation characteristic of the molecular vibration before and after collision.

This procedure is repeated at a series of reactant energies to give, ultimately, the complete excitation function and several other reaction details can often be deduced.

8.9 Calculation of Reaction Cross Sections

The most detailed calculations have been carried out[5] for the reaction

$$H + H_2 \rightarrow H_2 + H$$

using the Karplus and Porter semi-empirical potential energy surface described in section 5.4. After defining initial reactant conditions the next step is to calculate, using an appropriate averaging procedure, how reaction probability varies with impact parameter at the chosen reactant relative velocity.

[5] M. Karplus, R. N. Porter and R. D. Sharma, *J. Chem. Phys.*, **45**, 3871 (1966).
The results quoted in this section are taken from this reference.

Initial conditions

The first set of calculations is carried out at a fixed relative velocity. The internal energy states must also be specified—the potential energy surface is for ground electronic states and for the reactant molecule, H_2, the lowest vibrational and rotational states ($v = 0, J = 0$) are chosen.

The initial orientation of H_2 relative to the approach of the atom is usually given in polar coordinates R, χ, ϕ. It is also necessary to specify the internal momentum of H_2 by the angle of the momentum vector, η, relative to the perpendicular through the H_2 axis. All values of R, χ, ϕ, η are possible and to avoid an infinite number of calculations a representative average selection is made.

Trajectories are calculated for a selected impact parameter, b, each trajectory starting at an initial atom–molecule distance that is small enough to save computer time, but large enough to pick up the first significant $H \ldots H_2$ interaction.

Monte Carlo averaging

For ground state internal energies and selected relative velocity, the trajectory is calculated for a chosen impact parameter, but this trajectory encompasses the average result for R, χ, ϕ, η. As stated above, strictly speaking all possible values of R, χ, ϕ, η should be included, but to avoid exhausting computer time only selected initial values are covered. The distribution functions for these properties are well known, and the computer makes a random selection from these distributions in such a way that a fair representation of the distributions is maintained[6]. The process can be described as a weighted random selection of these initial parameters, and it has been termed, rather splendidly, *Monte Carlo averaging*.

Variation of reaction probability with *b*

At the selected reactant relative velocity and impact parameter, trajectories are calculated for various sets of initial parameters chosen by the Monte Carlo averaging procedure. If the total number of trajectories is $N(b)$ and $N_R(b)$ is the number that result in reaction, then the probability of reaction, $P(b)$, is:

$$P(b) = \underset{N \to \infty}{\text{Limit}} \left(\frac{N_R(b)}{N(b)} \right) \tag{8.9.1}$$

The limit is expressed as $N \to \infty$, but in practice calculations are continued until the probability is seen to tend to a clearly defined constant value. At the same relative velocity, this set of calculations is repeated for other impact parameters giving the variation of reaction probability or opacity function with b at fixed relative velocity as shown in figure 8.10.

Reaction cross section and excitation function

The contribution to the reaction cross section of collisions at impact parameters $b \to b + db$ is $P(b)2\pi b\, db$, where $P(b)$ is the reaction probability or opacity function at this value of b. Figure 8.10 shows how $P(b)$ varies from $b = 0$ to $b = b_{max}$, beyond which the probability is zero. Hence the cross section $S(v)$ at the selected velocity, v, is easily calculated from such data, since, from equation (8.4.2:) $S(v) = \int_0^{b_{max}} P(b)2\pi b\, db$.

[6] H. C. Blais and D. L. Bunker, *J. Chem. Phys.*, **37**, 2713 (1962).

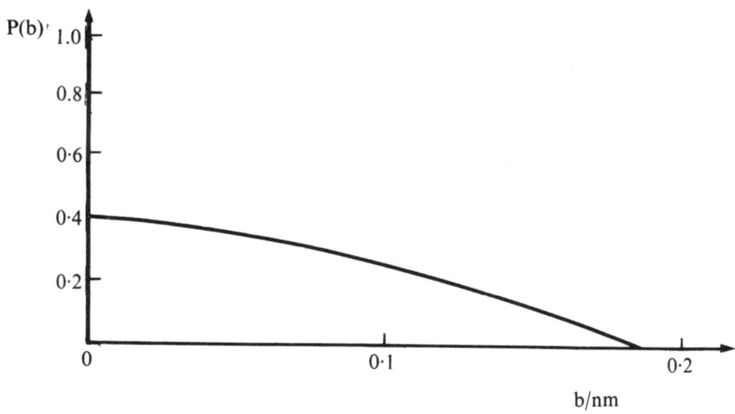

Figure 8.10
Variation of $P(b)$, the reaction probability (opacity function), with impact parameter, b, in the $H + H_2$ reaction where for H_2, $J = 0$, $v = 0$ and the relative velocity is $1{\cdot}17 \times 10^6$ cm s^{-1}.

The entire procedure thus leads to $S(v)$ at v and can be repeated for a series of velocities to give the excitation function. The result for the reaction between $H + H_2$ ($v = 0$, $J = 0$) is shown in figure 8.11, where relative velocity has been converted to *molar* translational energy, and the reaction cross section is thus expressed, $S(E)$.

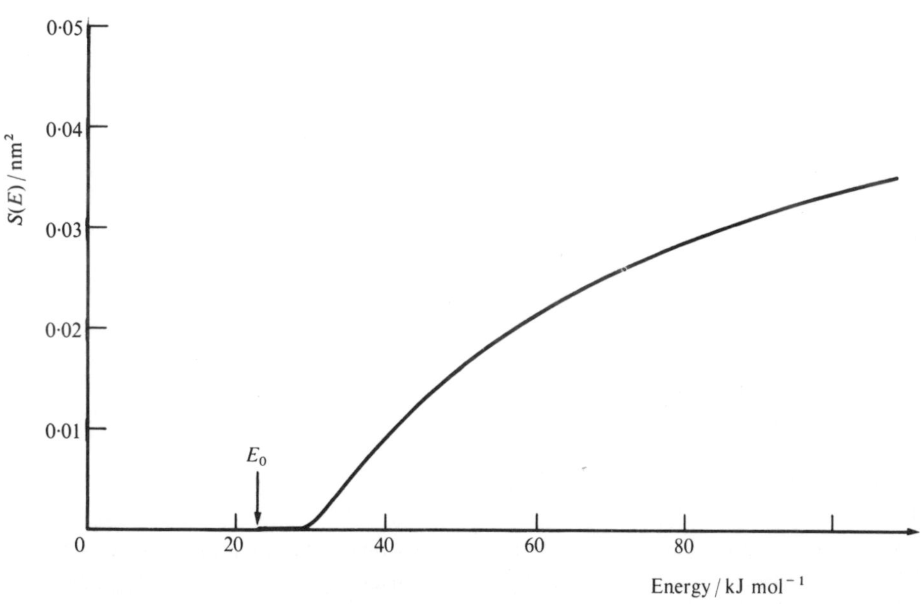

Figure 8.11
Variation of reaction cross section, $S(E)$, with relative translational energy for the $H + H_2$ reaction where for H_2, $J = 0$, $v = 0$.

The threshold energy, $E_0 = 23{\cdot}7$ kJ mol^{-1} but the cross section remains close to zero up to an 'effective' threshold of about 29 kJ mol^{-1}, whereupon it rises to an asymptotic value of 0·045 nm^2.

Other *J*, *v* values

In addition to the above calculations for the ground state reactant H_2, the effect of changing J up to $J = 5$ has been investigated, still with $v = 0$, but there is very little effect on the threshold energy or shape of the excitation function. The population of the $v = 1$ state of H_2 remains very low as far as 1000 K and this state has not been treated in detail. Some calculations indicate that the effect of the comparatively large vibrational quantum is a marked decrease in the threshold energy, which is not unexpected, as the considerable internal energy of the H_2 molecule can contribute to crossing the energy barrier to reaction.

8.10 Potential Energy Surfaces, Direct Interaction, Collision Complexes and Energy Distribution in Products

The success of molecular dynamical calculations depends on the validity of the potential energy surface chosen. The method has been illustrated for the system with the most reliable surface, but in less simple cases the theoretical construction of surfaces is much more difficult and must rely more and more on experiment for establishing some of the main features. Molecular beam and chemiluminescence experiments described in the next chapters reveal whether reaction is by direct, short-lived collision or if an intermediate collision complex is formed. In addition, for exothermic reactions the relative proportions of translational and vibrational excitation in the products can be measured. The potential energy surface can yield comparable results from the molecular dynamical calculations. A convenient example, studied in detail experimentally, is provided by the reaction:

$$K + CH_3I \rightarrow KI + CH_3$$

which is 92 kJ mol^{-1} exothermic. The methyl radical can be treated as an atom, reducing the system to the convenient triatomic type. Dynamical calculations have been carried out for several 'attractive' surfaces (see section 5.3) which differ in the details of short- and long-range interactions, and also for a surface with a potential energy basin[7].

On the attractive surfaces the trajectories were similar to those shown in figure 8.9. The collision time, the time for which the species are in close proximity, was very short, products separating within a vibrational period—almost the same as the time required for the atom to pass the molecule unimpeded. Thus the reaction proceeds by a direct or impulsive interaction. Experiment reveals an angular distribution of products clearly characteristic of such direct interaction, as discussed in section 9.4.

The attractive surfaces gave different but always high proportions of reactant exothermicity appearing as vibrational energy of product KI. However, the observed vibrational excitation in products suggests that the surfaces should have rather more repulsive character than those studied, a useful refinement for subsequent work.

[7] M. Karplus and L. M. Raff, *J. Chem. Phys.*, **41**, 1267 (1964); *J. Chem. Phys.*, **44**, 1212 (1966).

Highly exothermic reactions have very low threshold energies and the cross section does not vary greatly with reactant translational energy. The surfaces used gave maximum cross sections in the range 0·13–0·36 nm^2, the experimental value being about 0·35 nm^2.

The surface with the energy basin provides an instructive comparison. It results in reactive trajectories such as that shown in figure 8.12 which exhibits a long-lived collision complex. The complex persists for many vibrations and sometimes for several rotational periods. This is emphasized by comparison with the trajectory for direct interaction in figure 8.9. Calculations also indicate more equilibration between internal and translational product energies. It is shown in chapter 9 that this surface does not apply to the $K + CH_3I$ reaction, but other reactions with such features are described there. Hopefully such interaction between theory and experiment will lead to a growing body of reliable surfaces for wider ranges of reaction types.

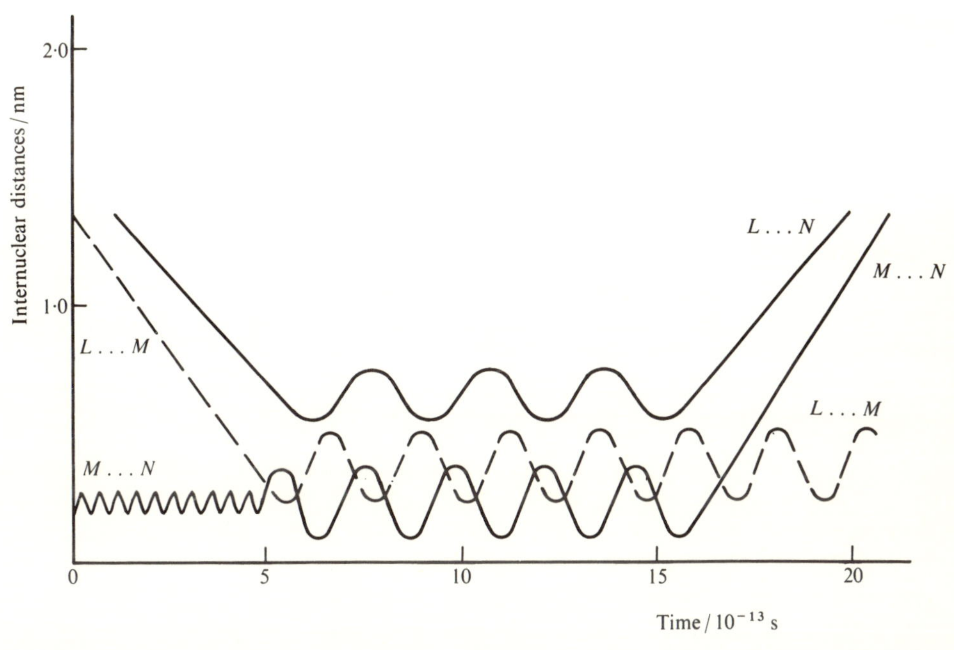

Figure 8.12
Variation of internuclear distance with time in collisions of L with MN where the reaction $L + MN \rightarrow LM + N$ occurs via a collision complex.

As L approaches MN and $L \ldots M$ and $L \ldots N$ distances decrease while $M \ldots N$ oscillates due to the molecular vibration. The three atoms remain in close proximity for a significant time, about 10^{-12} s in this example, undergoing several vibrations in the collision complex before N leaves the new molecule LM.

Recently a significant impetus has been given to such studies by success in the computation of potential energy surfaces for some few-electron cases such as $F + H_2$. Dynamical calculations based on such surfaces have predicted correctly such reaction characteristics as what type of reactant energy most efficiently promotes reaction, how reaction exothermicity appears in products, and isotope effects.

8.11 Calculated Rate Constants

When the excitation function for a reaction has been calculated the rate constant can be evaluated as described in section 8.6. The contribution of each rotational and vibrational state of reactant molecules should be included by summation over the appropriate distribution function. With the prototype $H + H_2$ reaction, for which the most reliable calculations have been made, the ground vibrational state makes by far the largest contribution in the temperature range 300–1000 K, and the effect of changing rotational state is relatively small at these temperatures. The results of molecular dynamical calculations are shown in table 8.1 where activated complex theory values and experimental results are given for comparison.

Table 8.1
Experimental Arrhenius parameters for rate constant of reaction: $H + H_2 \rightarrow H_2 + H$ compared with values calculated by molecular dynamical and activated complex theories.

	Pre-exponential factor (A)/ 10^{13} mol^{-1} cm^3 s^{-1}	Activation energy/ kJ mol^{-1}
Experimental	5·4	31·4 ± 4
ACT	7·4	37·2
Molecular dynamics	4·3	31·1

Thus, where a reliable surface is available, molecular dynamic theories can be extended successfully to the bulk kinetic properties expressed by the rate constant and activation energy. A future ideal is that, for few-electron problems at least, theoreticians will be able to start with the wave equation for the system and compute the reaction rate constant.

8.12 Energy Terms

There is a rather unfortunate proliferation of energy terms in experimental and theoretical kinetics. Experimental rate constant measurements give the *experimental activation energy*, E_{exp}, and molecular beam experiments measure the *threshold energy*, E_0, and the relation between them was examined in section 8.7. Activated complex theory and molecular dynamical calculations are based on classical potential energy surfaces (without zero-point energy) with their classical *energy barrier* to reaction—the difference between energy of reactants and the top of the barrier. Molecular dynamics yields the threshold energy but, as discussed in section 5.5, the appropriate energy barrier in activated complex theory is the difference between the zero-point energies of $H + H_2$ and the activated complex $H_3^‡$. This is shown as $E_0^‡$ in figure 8.13. The threshold energy is also shown on the diagram for comparison. This energy is sufficient for reactants to surmount the classical barrier, but insufficient to reach the fully quantized state of the activated complex—further indication of direct interaction without a definable molecular complex as intermediary.

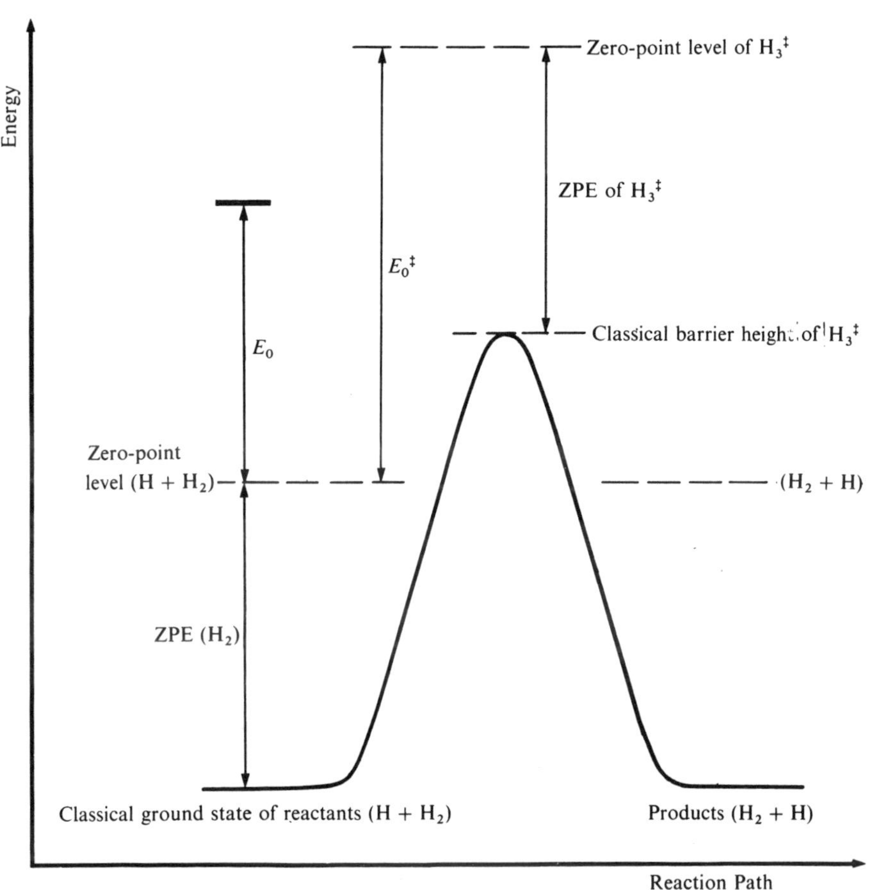

Figure 8.13
Reaction path for H + H_2 → H_2 + H, showing: the classical levels for reactants (H + H_2) activated complex ($H_3^‡$) and products (H_2 + H); their zero-point energies (ZPE); and the zero-point levels for reactants, activated complex and products. The zero-point energy of H_2 is quite large compared with the barrier height.

Also shown are the energy term for activated complex theory, $E_0^‡$, and the magnitude of the threshold energy, E_0.

Further Reading

E. F. Greene and A. Kupperman, *J. Chem. Ed.*, **45**, 361 (1968).

R. L. Wolfgang and A. Menzinger, *Angew Chem.*, **8**, 438 (1969).

K. J. Laidler, *Theories of Chemical Reaction Rates*, McGraw-Hill, 1969.

R. D. Levine and R. B. Bernstein, *Molecular Reaction Dynamics*, Oxford University Press, 1974.

J. C. Polanyi, *Accts Chem. Res.*, **5**, 161 (1972).

D. L. Bunker, *Accts Chem. Res.*, **7**, 195 (1974).

For comprehensive recent reviews:

R. D. Levine in *MTP Int. Rev. of Science, Phys. Chem. Ser. 1*, Vol. 1, (Ed. W. B. Brown), Butterworths, 1972.

R. A. Marcus, *Disc. Faraday Soc.*, **55**, 9 (1973).

Exercises

8.1
A molecular beam passes through a chamber, length 5 cm, containing inert gas at 5×10^{-5} Torr pressure, 300 K. The intensity of the beam is reduced by 20%. What is the total scattering cross section for molecule–inert gas collision?

(Ans: 2·77 nm^2)

8.2
The most probable relative velocity of the colliding species in the previous exercise was 1600 m s^{-1}. The total scattering cross section varies with relative velocity: $\sigma(v) = (\text{constant})\, v^{-1/2}$.

In a subsequent crossed-beam study, velocity selection of the molecular beam gave a most probable relative velocity of 400 m s^{-1}. The inert gas beam was 1 mm thick and the beam concentration was 3×10^{12} molecules cm^{-3}. What percentage of molecules was scattered from the beam?

(Ans: 1·7%)

8.3
The relationship between reaction rate constant and reaction cross sectional area is given by equation (8.6.2). Consider two simple models for the variation of reaction cross section with energy in the reaction $A + B \rightarrow$ products.

(a) $S(\varepsilon) = 0$ for $\varepsilon < \varepsilon_0$; $S(\varepsilon) = \pi\sigma_{AB}{}^2$ for $\varepsilon \geq \varepsilon_0$.

ε_0 is the reaction threshold energy. σ_{AB} is the sum of the 'hard-sphere' radii of A, B.

(b) $S(\varepsilon) = 0$ for $\varepsilon < \varepsilon_0$; $S(\varepsilon) = \pi\sigma_{AB}{}^2\,(1 - \varepsilon_0/\varepsilon)$ for $\varepsilon \geq \varepsilon_0$.

Show that the rate constant is given by

(a) $k = (\text{constant})\,(1 + \varepsilon_0/\bar{k}T)\,e^{-\varepsilon_0/\bar{k}T}$

(b) $k = (\text{constant})\,e^{-\varepsilon/\bar{k}T}$

where

$$\text{constant} = \pi\sigma_{AB}{}^2(8\bar{k}T/\pi\mu_{AB})^{1/2}$$

Compare the results with the SCT expression in chapter 5.

Make an approximate sketch of each excitation function, noting the limiting value of reaction cross section at high energy. Why does the fact that, for these simple models, the cross section does not fall to zero at very high energies introduce no serious error into the rate constant calculation?

N.B. $$\int_{x_1}^{x_2} x\,e^{-ax}\,dx = [-(ax\,e^{-ax} + e^{-ax})/a^2]_{x_1}^{x_2}$$

8.4
The relationship between activation energy and threshold energy is discussed in section 8.7. For the excitation function given in exercise 8.3, case (b), show that $E_{exp}/E_0 = 1 + RT/2E_0$.

Would the temperature variation for activation energy be experimentally detectable in the range 300–900 K where threshold energy is: (i) 20 kJ mol^{-1}; (ii) 80 kJ mol^{-1}?

8.5
For the reaction $H + H_2 \rightarrow H_2 + H$, the probability of reaction at a given relative velocity of reactants varies with impact parameter, b, as shown in figure 8.10. It was found that this curve is represented well by the function:

$$\text{Reaction probability} = a \cos(\pi b / 2b_{max})$$

What is the value of the reaction cross section when $a = 0{\cdot}39$, $b_{max} = 0{\cdot}181$ nm?

N.B. $$\int_{x_1}^{x_2} x \cos ax \, dx = [(ax \sin ax + \cos ax)/a^2]_{x_1}^{x_2}$$

(Ans: 0·0186 nm^2)

8.6
The cross sectional areas for the reaction: $H + H_2 \rightarrow H_2 + H$ at various relative translational energies are shown in the table:

E/kJ mol^{-1}	29	30	31	32	33	36	40	50	60	70	80	90
$S(E)$/nm^2 $\times 10^{-2}$	0·001	0·14	0·28	0·42	0·56	0·98	1·4	2·4	3·2	3·7	4·1	4·4

Calculate ($E\,e^{-E/RT}$) at these energies: (a) for $T = 300$ K; (b) at 900 K. Plot the functions $S(E)$ and $S(E)E\,e^{-E/RT}$ against E at each temperature. Evaluate graphically the ratio of rate constant values at the two temperatures and hence calculate the activation energy.

(Ans: 31 kJ mol^{-1})

Chapter 9: Molecular Beams

Molecular beams are dilute enough not to become diffuse on passing through regions of very low background pressure. They have been used in the physical sciences for many years, but were first applied to reaction kinetics only twenty years ago with the development of a detector sensitive enough to cope with very low product concentrations.

When two molecular beams cross, any species will undergo only one collision in the crossing zone. The result may be elastic, inelastic or reactive scattering away from the direction of the incident beams but our discussion is limited mainly to reactive collisions and detecting scattered reaction products. Beam experiments have become very sophisticated and, with appropriate additions to beam sources, the relative translational energy of reactants and perhaps their internal energy states can be defined with some precision. For each reactant energy, products are detected emerging from the crossing zone at different angles and the reaction cross section can be evaluated from total product yields. Varying the reactant relative translational energy gives the cross section as a function of energy and hence the threshold energy. Techniques have also been developed to measure product velocities at selected angles, which allows translational and internal product energies to be calculated. This leads to further deductions such as whether the collisional interaction is short-lived or if an intermediate collision complex is formed, and the main features of the potential energy surface for the reaction can sometimes be constructed.

9.1 Experimental[1]

The main components of the apparatus are shown schematically in figure 9.1. They are: the beam sources, S; velocity selector for reactants, A; detector, D; and velocity analyser for products, B. The apparatus is contained in a chamber at very high vacuum. B and D are movable for investigation of products at different angles. These major components will be described below and some further refinements included.

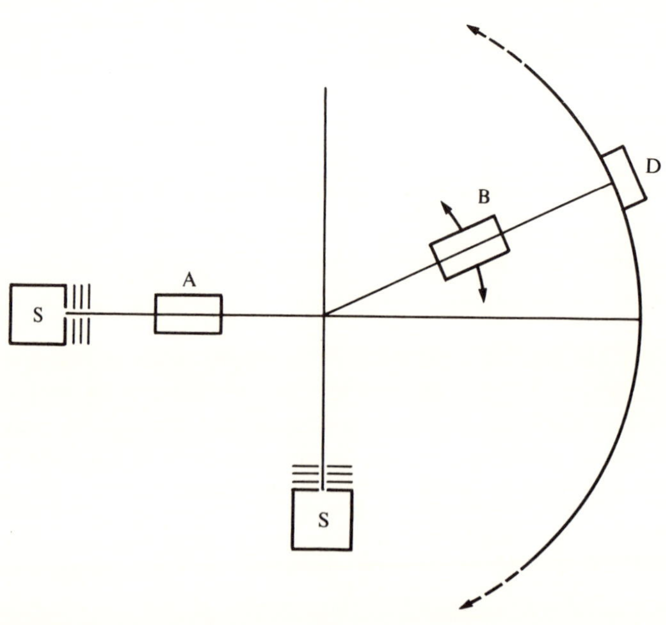

Figure 9.1
Schematic drawing of main components of crossed molecular beam apparatus. S, beam sources; A, reactant velocity selector; B, product velocity analyser; D, detector.

Beam sources

The most common sources have been small *ovens*, from which the gas escapes through a small slit into the high-vacuum chamber. To ensure that molecules undergo no collisions in passing through the slit their mean free path must be greater than the slit width, so the pressure in the oven must be low, say 0·1 Torr, 13 N m^{-2}. This gives effusive flow and the molecules in the beam travel in straight line continuations of their original trajectories. A series of slits then collimates the molecules into a narrow beam. The velocity distribution in the beam is governed by oven temperature and if a narrow range of translational energies is required a velocity selector must be added. Beam intensities are low but can be increased to some extent by multichannel oven exits.

[1] Recent reviews include D. R. Herschbach, *Disc. Faraday Soc.*, **55**, 233 (1973); J. L. Kinsey in *MTP Int. Rev. Sci., Phys. Chem. Ser.* 1, Vol. 9, (Ed. J. C. Polanyi), Butterworths, 1972.

Greater beam intensities and higher translational energies can now be attained with *supersonic nozzle sources*. The oven source is replaced by a low density supersonic jet produced when a gas undergoes free hydrodynamic expansion through an appropriate nozzle. The gas stream is directed onto a small orifice or 'skimmer', and if the velocity of the gas jet is large relative to the thermal molecular velocities the molecules passing through the skimmer are focused on the beam axis. The beam intensity is high and the molecules can have a high mean velocity in the beam direction with a very narrow range of velocities about the mean. This gives well-defined translational energy for the species, and energies much higher than those possible with oven sources can be attained.

In a further extension of supersonic sources, a beam of non-reactive species can be 'seeded' with reactants added at low relative concentration.

Velocity selection

Reactants leave *oven* sources with Maxwellian velocity distribution, but a narrow velocity range can be obtained using a device which rejects species with velocities higher or lower than that required. This *velocity selector*, in its simplest form, consists of a series of discs mounted on a rotating axis which is parallel to the beam direction. Each disc has a slot through which part of the beam passes and successive discs are offset so that species with the required velocity pass straight through whereas those with higher or lower velocity miss the slots and are deflected off the rotating disc. The velocity obtained is determined by the speed of rotation which is controlled easily, making selection of reactant translational energy simple.

The great disadvantage of this method is that it reduces beam intensity and consequently reduces product intensity at the detector. To ensure sufficient product intensity, early studies were of exothermic reactions with very low threshold energies and quite large cross sections. Supersonic and seeded beams have well-defined velocity and require no further selection. They provide greater intensities and much higher reactant velocities and it is possible to investigate reactions at much larger energies, even endothermic reactions with high thresholds.

If one beam is velocity selected, quite well-defined relative velocities are obtained without imposing velocity selection on the second beam—which would reduce product intensity still further.

Internal energy of reactants

Only ground electronic and vibrational states are appreciably populated in most cases but some experiments have been carried out with reactant molecules in an excited vibrational state produced, for example, by irradiating the reactant beam with suitable laser light.

Molecules in the beam have equilibrium rotational energy distribution, but it is possible to select rotational states of polar molecules using focused electrical fields which deflect unwanted states out of the beam direction. However, the consequent reduction in beam intensity is such that it is far more difficult to study the effect of rotational states on reactivity.

Orientating reactants to study steric effects

If a reactant is polar and exhibits a first-order Stark effect (e.g. CH_3I), passage through a six-pole electric field gives a beam orientated so that the other reactant approaches predominantly one end of the molecule. With opposite setting of the polarizer the reactant approaches the other end of the molecule and the ratio of reaction cross sections found for the two cases will measure a steric effect for the reaction.

Beam chamber and background pressures

Molecular beams are very dilute and background pressure must be minimized to achieve maximum sensitivity for product detection. The necessary background level depends on the particular reaction and detector, and successive differential pumping with cryogenic traps is required. Typical beams have reactant pressures of about 10^{-5} Torr, $1{\cdot}3 \times 10^{-3}$ N m^{-2}, in the crossing zone, and with a reaction cross section of 0·01 nm^2, the products are at approximately 10^{-10} Torr, $1{\cdot}3 \times 10^{-8}$ N m^{-2}, in the crossing zone and several orders of magnitude less at the detector.

Detector

When dilute beams cross, only a small fraction of the species collides. A high proportion of collisions results in elastic or inelastic energy transfer and the yield of reactive collisions is very small. Furthermore, products are scattered through a wide angular range and the product intensity within a narrow solid angle is very low—expressed as concentration near the detector it is perhaps only 10^2 molecules cm^{-3}, or about 10^3 molecules cm^{-2} hitting the detector every second.

It is not surprising that detector sensitivity has long been a major limiting factor in molecular beam research. Modern beam reaction kinetics started in 1954 when Taylor and Datz[2] discovered 'a simple two filament (one clean, one dirty!) *surface ionization detector*'. An alkali metal atom striking a hot tungsten filament ionizes, giving an ion current that can be detected with a sensitive ammeter but unfortunately alkali metals and alkali halides have similar ionization efficiencies. However, a filament of platinum alloyed with a small percentage of tungsten has a much higher efficiency for ionizing alkali metals than for the halides, and a combined system of both filaments gives a detector suitable for reactions of the type

$$M + AX \rightarrow MX + A$$

since one detector measures metal atom plus halide, ($[M] + [MX]$), the other just the atom $[M]$, giving the product $[MX]$ by difference.

Many alkali metal atom reactions were investigated with this extremely sensitive detector before the next major advance was made about a decade later. This was the arrival of the '*universal detector*'—electron impact ionization of the product, with mass spectrometric detection of the resulting ion. Such techniques had been thought insufficiently sensitive but, in general, they bear comparison with surface ionization. The pessimistic view of mass spectrometric detection arose because early efforts failed to reduce the interfering background sufficiently. This was overcome and then extra sensitivity gained by beam modulation for improving the signal: noise ratio.

[2] E. H. Taylor and S. Datz, *J. Chem. Phys.*, **23**, 1711 (1955).

The great importance of the universal detector, the reason it has come to dominate this field of study, is that it extends beam methods beyond the constraints of alkali metal systems to, for example, reactions of hydrogen atoms, halogen atoms and alkyl radicals.

Velocity analysis of products

The time-of-flight mass spectrometer gives a convenient method for comprehensive velocity analysis of products. At a chosen product angle, a pulse of products is admitted and the intensity measured as a function of flight time down the spectrometer, which is directly related to product velocity. Even with signals of the order of 10 counts s^{-1} at the peak of the velocity distribution, a complete record over the whole range of angles can be obtained in only a few hours. The combination of mass spectrometric detection and velocity analysis is now predominant in molecular beam studies.

Internal energy of products

The velocity of products gives their translational energy, and if the heat of reaction is known their internal energy can be calculated from energy conservation. Product vibrational and rotational states can sometimes be detected by irradiating products with appropriate laser light and observing fluorescence from the excited states formed. Other methods include deflection in an inhomogeneous field to obtain rotational energies of polar molecules and electron resonance spectroscopy to detect vibrational population.

9.2 Range of Reactions Investigated

With the recent addition of supersonic beam sources and universal mass spectrometric detectors molecular beam experiments have multiplied in number and scope. Herschbach has described the early work on alkali metal atom reactions, using oven sources and surface ionization detectors, as the 'early alkali age', and the increasingly sophisticated investigations of lithium atoms, diatomic alkali molecules and alkali halide salts as the 'middle alkali age'. Most current studies concentrate on atom transfer reactions of hydrogen atoms, alkaline earth atoms, oxygen atoms and methyl radicals. Reactions of halogen atoms with olefin and aromatic hydrocarbons are under investigation on a large scale and should lead to the 'organic age'. Some indication of range of these applications is given in simplified form in table 9.1.

Table 9.1
Some types of reaction investigated by molecular beam techniques. $M \equiv$ alkali or alkali earth metal atom; $\mathrm{H} \equiv$ hydrogen or its isotopes; $X \equiv$ halogen atom; $R \equiv$ alkyl radical.

$M + \mathrm{H}X \rightarrow MX + \mathrm{H}$	$X + \mathrm{H}_2 \rightarrow \mathrm{H}X + \mathrm{H}$
$M + X_2 \rightarrow MX + X$	$X + \mathrm{H}X \rightarrow \mathrm{H}X + X$
$M + RX \rightarrow MX + R$	$X + X_2 \rightarrow X_2 + X$
$M + MX \rightarrow MX + M$	$X + \mathrm{C_2H_4} \rightarrow \mathrm{C_2H_3}X + \mathrm{H}$
$M + \mathrm{O}_2 \rightarrow M\mathrm{O} + \mathrm{O}$	$\rightarrow \mathrm{C_2H_3} + \mathrm{H}X$
$\mathrm{H} + \mathrm{H}_2 \rightarrow \mathrm{H}_2 + \mathrm{H}$	$\mathrm{O} + X_2 \rightarrow \mathrm{O}X + X$
$\mathrm{H} + \mathrm{H}X \rightarrow \mathrm{H}X + \mathrm{H}$	$R + X_2 \rightarrow RX + X$
$\mathrm{H} + X_2 \rightarrow \mathrm{H}X + X$	
$\mathrm{H} + M_2 \rightarrow M\mathrm{H} + M$	

In the next section, one reaction is chosen to illustrate the type of information that can be obtained by the molecular beam method, and this will be followed by a general discussion of some of the important topics that arise.

9.3 The Reaction $K + CH_3I \rightarrow KI + CH_3$, Reactive Scattering and Cross Section

This reaction has been the subject of more than thirty experimental and theoretical papers from the early alkali age to the present day[3]. A typical beam arrangement for study of angular scattering is shown in figure 9.2. The potassium and methyl iodide beams cross at right angles and the detector measures the intensity of species scattered at laboratory angle α.

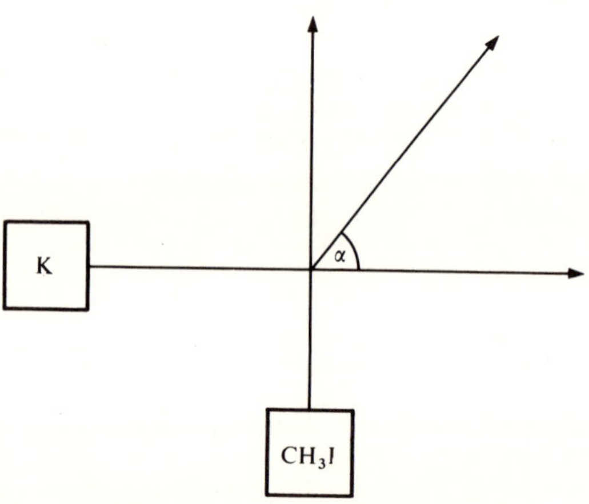

Figure 9.2
Schematic representation of K and CH_3I beams crossing at right angles, and of laboratory scattering angle α.

One of the earliest studies used only a potassium atom surface ionization detector and the relative translational energy of reactants was defined by velocity selection of the potassium beam. Most of the scattered atoms result from elastic collisions, but at large angles the intensity of atoms was less than expected from elastic scattering measurements in systems of similar size, such as K + Xe. This loss of intensity for potassium atoms was attributed to reactive collisions forming KI, which was not detected in this experiment. From the missing intensity at any angle the differential cross section at that angle is calculable in absolute units, and summation over all angles gave the reaction cross section—from equation (8.3.3):

$$S(E) = \int 2\pi\{S(E, \theta)_{EL} - S(E, \theta)_{obs}\}\sin\theta \, d\theta$$

[3] Results have been reviewed in E. F. Greene and J. Ross, *Science*, **159**, 587 (1968); R. B. Bernstein and A. M. Rulis, *Disc. Faraday Soc.*, **55**, 293 (1973).

The experiment was repeated at several reactant energies and the result, the excitation function for this reaction, is shown in figure 9.3.

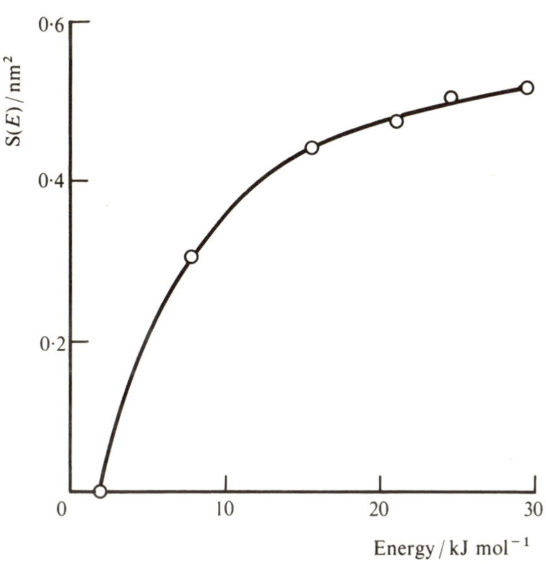

Figure 9.3
The excitation function for the reaction $K + CH_3I \rightarrow KI + CH_3$ estimated from non-reactive scattering measurements.

Another approach to measuring angular scattering is to use two detectors, one of which is sensitive to K, the other to K + KI. The first experiments of this type did not employ velocity selection, the relative reactant translation energy being taken from the most probable reactant velocities. A slightly simplified version of the results for this energy is shown in figure 9.4. The potassium iodide formed at the various angles is given by the difference between the two curves, and the resulting product angular distribution is shown in figure 9.5. Even this fairly crude example of a molecular beam experiment gave the extremely important result that the angular distribution was *anisotropic*, the product was scattered over a comparatively narrow range of angles with a quite well defined most probable angle, $\alpha_p = 83°$. From appropriate summation of the potassium iodide intensity it was possible to calculate the reaction cross section at this energy, and this gave good agreement with the previous results.

Later experiments with a seeded supersonic methyl iodide beam yielded detailed angular and velocity distributions for products over a wider range of energies. Angular distributions are easier to obtain than *absolute* measurements of product intensities, and it is easier to find the shape of the excitation function than the *absolute* value of the cross section. The experimental excitation function is shown in figure 9.6[4]. The absolute cross sections were obtained by assuming agreement with previous experiments

[4] M. Gersh and R. B. Bernstein, *J. Chem. Phys.*, **56**, 6131 (1971).

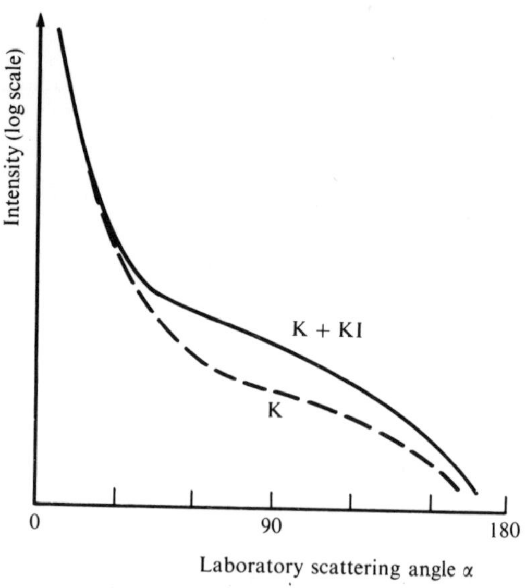

Figure 9.4
Angular distribution of: (a) K alone; and (b) K + KI scattered on the interaction of thermal K and CH_3I beams.

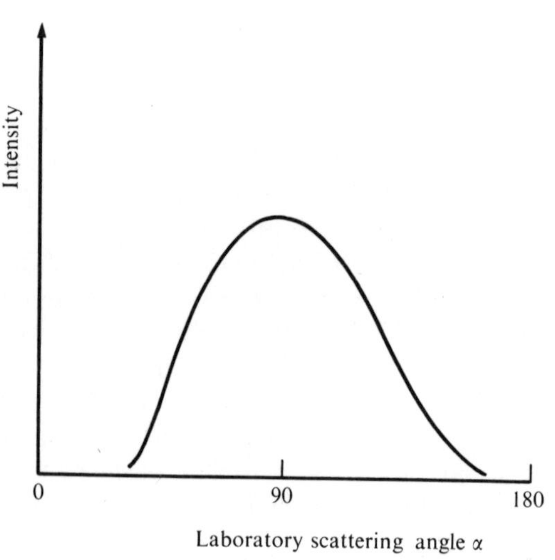

Figure 9.5
Angular distribution of KI from the interaction of thermal K and CH_3I beams.

(see figure 9.3) where cross sections were deduced from non-reactive scattering data, taking all non-reactive scattering to be elastic. The results differ above about 20 kJ mol^{-1} probably due to a significant inelastic contribution to non-reactive scattering at these energies.

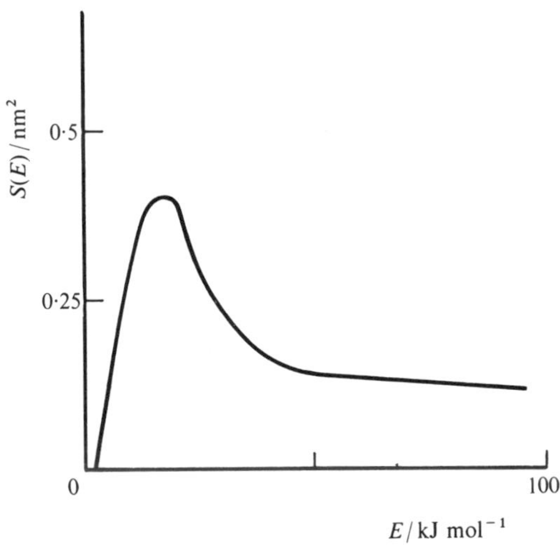

Figure 9.6
Excitation function for K + CH_3I reaction, from direct measurement of KI product. The results did not give absolute values of $S(E)$, and the curve is scaled to fit the best early results at low energy.

9.4 The Collision in Centre of Mass Coordinates: Forward and Backward Scattering

Some of the salient features of molecular collisions emerge with greater clarity when events are described in a coordinate system moving with the centre of mass (c.m.)—when velocities of reactants and products are expressed relative to the c.m. The results for laboratory scattering angle are similarly transformed to scattering angle in c.m. coordinates. The mathematical complexities of these procedures are discussed elsewhere[5], but though the full calculations may be difficult there are some very convenient diagrammatic representations.

Velocity is a vector quantity, it has magnitude and direction and can be represented by a line in the appropriate direction whose length is proportional to its magnitude. The relative velocity of two species is given by a simple vector triangle. Consider, for example, the usual experimental arrangement of a beam of species A crossing a beam of BC at right angles. The velocities in laboratory coordinates, v_A and v_{BC}, can be drawn as vectors of length l0 and n0 as in figure 9.7(a). The relative velocity, v, is then given by the vector ln that completes the triangle. The diagram can also be drawn with the vectors for A and BC shifted to a common origin, 0, as in figure 9.7(b).

The two species have a centre of mass which can be defined as the point where the total mass of the system acts or appears to act. If the species have masses m_A and m_{BC} acting at points l and n then the c.m. is at point p where: $m_A \times \mathrm{lp} = m_{BC} \times \mathrm{np}$. In the example shown in figure 9.7(a), when A and BC move from points l and n to 0,

[5] J. L. Kinsey in *MTP Int. Rev. Sci., Phys. Chem. Ser. 1*, Vol. 9, (Ed. J. C. Polanyi), Butterworths, 1972.

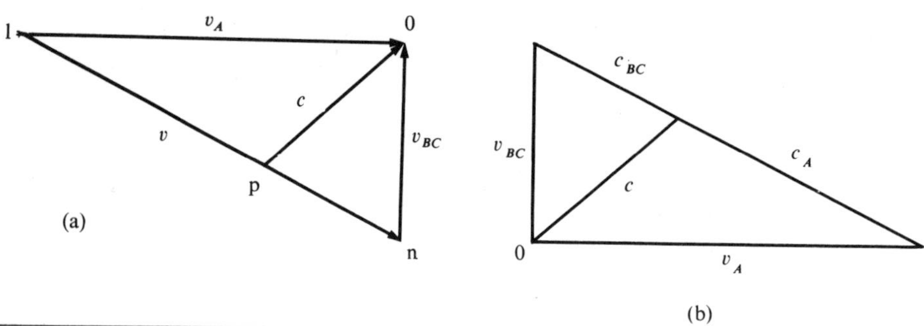

Figure 9.7
(a) Velocity vector diagram for beams intersecting at right angles. The velocity vectors are v_A, v_{BC}, v is the relative velocity and c the centre of mass velocity vector.

(b) Vectors are shifted to a common origin and centre of mass velocities c_A, c_{BC} are shown. This is a Newton diagram.

the c.m. moves from p to 0. p0 gives the direction and magnitude of c, the c.m. velocity, that is p0 represents the c.m. velocity vector.

The velocities of A and BC can be expressed relative to the c.m. velocity. An 'observer' situated at the c.m. 'sees' A and BC approach him from opposite directions, and the velocities with which they approach, their velocities in the c.m. system, are portrayed very simply on the vector diagram. The velocity of A relative to the c.m., c_A, is given by 1p and that for BC, c_{BC}, by np in figure 9.7(a). c_A and c_{BC} are also indicated on figure 9.7(b). The diagrams also show that the relative velocity, v, is related to the velocities in the c.m. system by : $v = c_A - c_{BC}$. The sign indicates the opposite direction of the vectors.

Diagrams with the laboratory velocity vectors shifted to a common origin as in figure 9.7(b) are called Newton diagrams and they provide a convenient summary of the restrictions imposed by kinematics—conservation of mass, energy, and momentum.

Elastic scattering

Most collisions in gases result in *elastic scattering* where the magnitude of the relative velocity vector is the same before and after collision but its direction may change. The velocity of the c.m. is also unaffected by the collision. A and BC are scattered and the angle between the relative velocity vectors before and after collision, θ, is the scattering angle in the c.m. system. In fact, elastic scattering at each θ can give a cone of scattered species about the initial relative velocity vector, a cone that intersects the plane of the apparatus twice. This situation is shown in figure 9.8 where the Newton diagram before collision is indicated, for reference, by dashed lines, and the intersection of the cone with the experimental plane is along the two lines XX′ and YY′. The angle θ is shown for scattering of BC. As stated above it is the angle between the initial relative velocity vector (ZZ′) and the final relative velocity vector (XX′). For c.m. scattering angle θ there will be two laboratory scattering angles, which, for BC, are defined by the laboratory velocity vectors after collision. These are not drawn on the diagram, but will be given by the angle between the incident A direction, O_LZ', and the directions after collision, O_LX and O_LY. The corresponding laboratory angles α_{BC} are drawn for these two cases in figures 9.9(a) and (b), the Newton diagrams for the scattered species. The laboratory scattering of A through two angles α_A is also shown and the laboratory velocity vectors are labelled v_{BC}' and v_A' on the two diagrams.

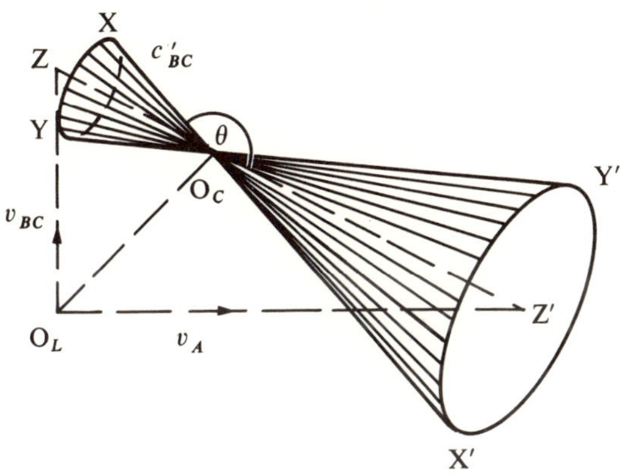

Figure 9.8
Elastic scattering $A + BC \rightarrow A + BC$.

The Newton diagram before collision is drawn in dashed lines. The initial velocity vectors are v_A (O_LZ'), v_{BC} (O_LZ) and the relative velocity is given by ZZ'.

The centre of mass scattering angle, θ, is the angle between relative velocity vectors before and after scattering. One centre of mass angle θ gives a cone of possible scattering and the cone intersects the plane of the apparatus along XX' and YY'. Thus for the same θ, BC is scattered along O_IX and O_LY in the experimental plane.

The corresponding laboratory velocity vectors for BC are O_LX in one case, and O_LY in the other. These lines have not been drawn in this figure but are included in the Newton diagrams for the two situations after collision, figures 9.9(a) and 9.9(b). The centre of mass velocity after collision, c_{BC}', is given by O_CX or O_CY.

The centre of mass velocity vectors for A after collision are given by O_CX' and O_CY', the laboratory vectors are O_LX' and O_LY'.

Reactive scattering (backward)

Reactive scattering embraces an extra complication due to the energy changes in reaction. In most beam studies the reaction has been exothermic and some part of the exothermicity may appear as translational energy of products, giving them a greater relative velocity than the reactants possessed. Consider the reaction

$$A + BC \rightarrow AB + C$$

where the velocity vectors before collision are v_A and v_{BC}. These are shown in the Newton diagram for reactants (the dashed lines) in figure 9.10, which also portrays the *most probable* c.m. scattering angle θ. As explained above, this gives two laboratory scattering angles, one of which, α, is shown in figure 9.10. The laboratory velocity of

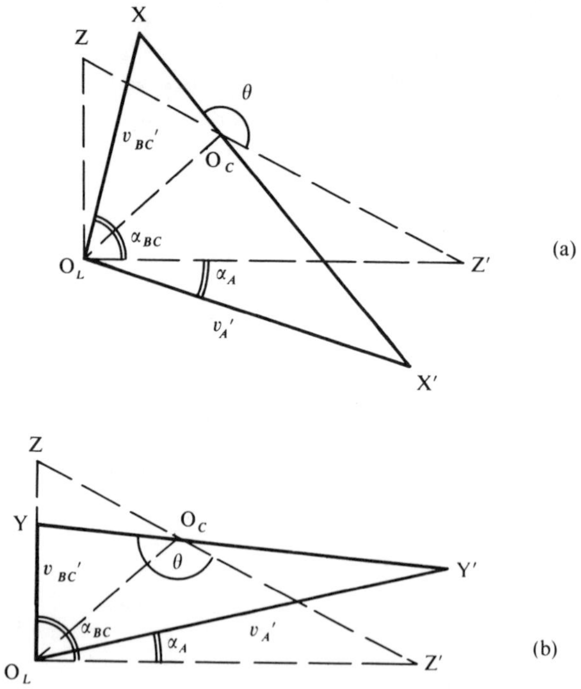

Figure 9.9
Elastic scattering $A + BC \rightarrow A + BC$.

Newton diagrams after collision, the diagram before collision is drawn in dashed lines. Scattering at centre of mass angle θ gives a cone that intersects the experimental plane twice, for relative velocity vector XX′ as in (a) and YY′ as in (b).

(a) After collision the centre of mass velocities for BC and A are O_CX and O_CX' respectively and the corresponding laboratory velocities are v_{BC}' (O_LX) and v_A' (O_LX'). The laboratory scattering angles, measured from the direction of the incident A beam, are α_{BC} and α_A.

(b) After collision the centre of mass velocities are O_CY and O_CY', the laboratory velocities v_{BC}' (O_LY) and v_A' (O_LY') and the laboratory scattering angles are α_{BC} and α_A.

product AB is shown as v_{AB}'. In c.m. coordinates the product AB has recoiled in direction O_CX at scattering angle θ and its c.m. velocity, c_{AB}' is given by this length O_CX. The vector, v_C', for product C is drawn to complete the Newton diagram, and its c.m. velocity, c_C', is O_CX' as shown.

A simplified c.m. representation of the event can be extracted from the Newton diagram showing only the initial approach of A to BC and the separation of products AB and C in the c.m. system. These are redrawn in figure 9.11 and labelled appropriately, though the lines here are not proportional to velocity. The c.m. scattering angle is shown, and θ is close to 180°. Clearly AB is scattered backwards to the direction from which A approached, it *rebounds* from collision and this is called *backward scattering*. This is the case for the most probable scattering in the $K + CH_3I$ reaction which corresponds closely to figures 9.10 and 9.11 where $K \equiv A$ and $CH_3I \equiv BC$.

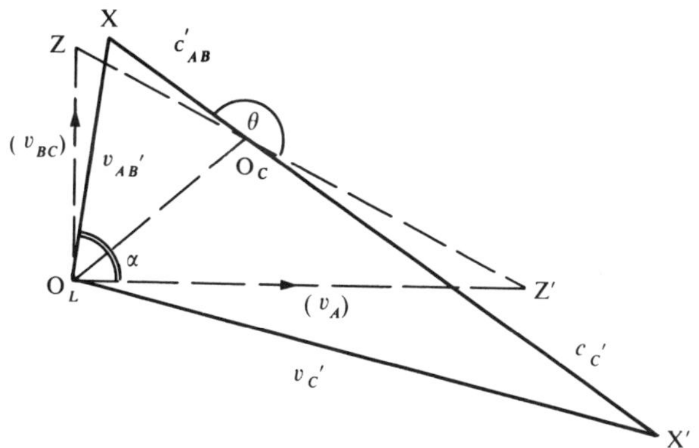

Figure 9.10
The reaction $A + BC \rightarrow AB + C$.

The reaction is exothermic, some energy appears as translational energy of products giving a greater relative velocity vector (XX′) then for reactants (ZZ′).

The Newton diagram for A and BC before collision is drawn in dashed lines. After collision the product AB is scattered through centre of mass angle θ, giving laboratory angle α. There is another α corresponding to the cone of scattering at this θ, but it is omitted for clarity. The centre of mass velocity vectors are c_{AB}' (O_CX) and c_C' (O_CX') and laboratory velocity vectors are v_{AB}' (O_LX) and v_C' (O_LX').

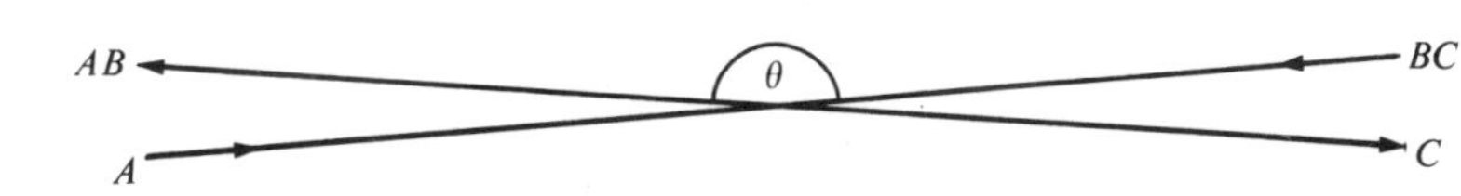

Figure 9.11
The reaction $A + BC \rightarrow AB + C$ in the centre of mass system showing backward scattering with centre of mass scattering angle θ close to 180°. The lines are *not* proportional to velocity.

Detailed product velocity distribution

The results of early experiments on reactions with predominantly backward scattering have been refined greatly by the latest techniques for measuring product velocity distribution at all scattering angles. These results are neatly summarized by a diagram such as figure 9.12 which was obtained for $K + CH_3I$ at a single reactant energy. The contours show the relative probability of KI appearing at a given angle and velocity†—the velocity is proportional to the distance from the c.m. at the origin. The angles and velocities of highest probability are emphasized by shading. The most probable angle

† Nearly monoenergetic reactant beams and careful velocity analysis of products is necessary to prepare such diagrams, since product intensity at a given laboratory angle contains contributions from scattering at different product translational energies and c.m. angles.

is close to 180°, and the most probable translational energy calculated from the velocity is about 70 kJ mol^{-1} for a reactant energy of 12 kJ mol^{-1}. With a reaction of known exothermicity this result makes it possible to calculate the most probable internal energy for products.

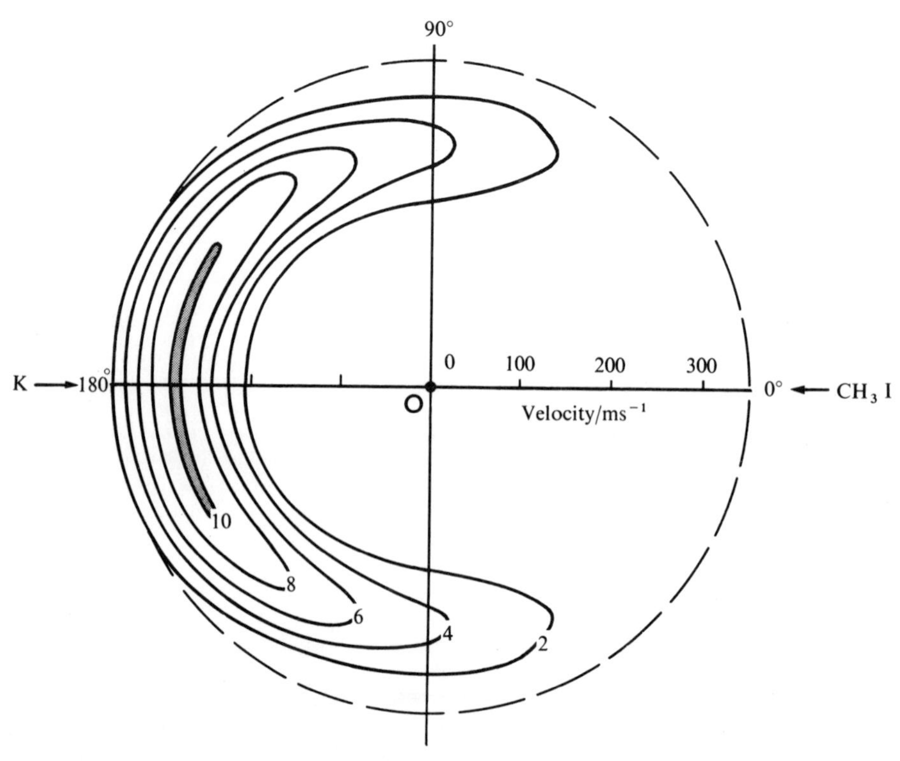

Figure 9.12
The reaction $K + CH_3I \rightarrow KI + CH_3$.

KI product angular and velocity distribution diagram at one relative reactant energy (~12 kJ mol^{-1}). In the centre of mass system reactants approach the centre of mass (the origin O) from opposite directions. The centre of mass scattering angles vary from 0–180° on both sides of the line of approach. The velocity is proportional to distance from the origin and the velocity scale is shown on one axis.

The contours represent increasing probability of products scattered at the angle and velocity indicated. The maximum probability, highlighted by shading, occurs close to 180°—backward scattering. The dashed line shows the maximum product velocity calculated from energy conservation.

Forward scattering

Backward scattering is not the only type of anisotropy found in the angular distribution of products. Many reactions show a peak yield at c.m. angle, $\theta = 0°$. For detailed product velocity and angular distribution experiments the result would resemble those shown in figure 9.12, but reversed so that the maximum probability appeared around $\theta = 0$. Another convenient way of representing such collisions is in a diagram analogous to figure 9.11. A typical example is given in figure 9.13, which shows that the product AB continues in much the same direction that reactant A approached the c.m. This is *forward scattering*. A appears to *strip* B from BC, and proceed in almost the same direction.

Figure 9.13
The reaction $A + BC \rightarrow AB + C$ in the centre of mass system showing forward scattering, with centre of mass scattering angle θ close to 0°. The lines are *not* proportional to velocity.

This covers the two extreme cases of anisotropic angular distributions, but many intermediate cases have been identified. In addition, symmetrical product distributions are sometimes found and these, as discussed in section 9.7, are characteristic of relatively long-lived collision complexes.

9.5 Further Details for K + CH_3I

Duration of collision

If, when reactants collide, they form a complex undergoing several rotations, this would be equally likely to decompose in any phase of rotation giving equal signals forward and back with respect to the centre of mass. Anisotropic product scattering implies that the duration of the reactive collision must be very short. Thus the lifetime of any K..I..CH_3 collision complex must be less than a period of rotation, say less than 10^{-12} s, and this reaction proceeds through a direct or impulsive interaction.

Distribution of energy in products

For an exothermic reaction:

$$[E_{trans} + E_{int}]_{reactant} = [E_{trans} + E_{int}]_{products} + \Delta E$$

where ΔE is the energy change for reaction, in this example, -92 kJ mol^{-1}.

Under experimental conditions reactants are in the ground electronic state and predominately the ground vibrational state. Thus the small rotational energy, around 2 kJ mol^{-1}, is the total internal energy of reactants. Typical reactant translational energies are in the range 4–20 kJ mol^{-1} and so 96–112 kJ mol^{-1} is partitioned between internal and translational energy of the products. The experimental analysis of product velocities thus allows a direct assessment of the energy partitioning and for this reaction the fraction of energy appearing as vibration appears to be 50% or less. Some recent experiments with a surface ionization detector that is sensitive to vibrational excitation show that the potassium iodide product is formed in high vibrational levels, and so the methyl radical is relatively unexcited.

Such observations contribute greatly to understanding the potential energy surface for the reaction as discussed in section 8.10.

Steric effect

The reactant CH_3I is a polar molecule and it is possible, using a six-pole electric field, to impose an orientation on the methyl iodide so that the potassium atom predominately approaches the iodine end. Reversing the field then results in the atom mainly approaching the methyl end. These experiments confirmed that reaction was more probable when the atom approached the iodine end. Some interesting effects are found for other reactions, however (see section 9.9).

Summary for K + CH_3I

Molecular beam experiments give a picture of an exothermic reaction with low threshold energy proceeding by direct collisional interaction which gives mainly backward scattering of products. Somewhat less than half the exothermicity appears as internal energy of the products, mostly in KI. There is a clear steric effect. Since most of this information is quantitative we are presented with an outstanding extension of kinetic knowledge to the most fundamental level yet attained.

The picture is not complete. The correlation between results where products are observed directly and where information is deduced from non-reactive scattering is not yet satisfactory, and the importance of the polyatomic character of the methyl radical has not been elucidated. The success achieved so far whets the appetite for such information as: what happens at energies high enough for competing processes such as collision-induced dissociation; and what would be the effect of internal excitation of methyl iodide? Furthermore, so much experimental information is now available that theoretical models, formerly adequate, no longer come close enough to simulating the details[6].

The more important features of reactive collisions that emerged in this example will now be examined for the general types of reaction that have been investigated by molecular beam methods.

9.6 Direct Interaction: Rebound and Stripping

Most reactions with anisotropic product distribution can be divided into two types, those with mainly forward, and those with mainly backward scattering. The reaction already discussed in detail, K + CH_3I, belongs to the latter category and is designated a *rebound reaction*. Other examples of this type are: the reaction of potassium with other alkyl halides; analogous reactions of sodium; reactions of hydrogen or deuterium atoms with halogens; and methyl radical–halogen reactions. Rebound processes generally have quite low cross sections and a relatively high fraction of their exothermicity appears as translational energy—indicating potential energy surfaces of largely repulsive character. An oversimplified but useful insight into this mechanism is provided by considering the opacity function, how reaction probability varies with impact parameter, b. Here the reaction only occurs for small impact parameter

[6] R. B. Bernstein and A. M. Rulis, *Disc. Faraday Soc.*, **55**, 293 (1973).

collisions. The re-arrangement of reactants to products takes place within close proximity and newly formed products separate under the influence of short-range repulsion. So these direct reactions, occurring mainly at small b values, have small reaction cross sections and backward scattering of products.

Stripping reactions are those with forward product scattering. They have large cross sections ($>1\ nm^2$) and only a small fraction of product energy is translational, a property usually associated with attractive potential energy surfaces. This category of reactions has been more widely investigated than any other and includes the reactions of alkali metal atoms with halogens, interhalogens, BrCN, PBr_3, PCl_3, and $SnCl_4$. Stripping reactions must also have rebound components due to nearly 'head-on', low impact parameter collisions. However the reactions have more important long-range contributions due to interaction at large b values. In large b encounters the reactant atom carries forward the second atom picked up from the other reactant, with very little consequent deflection. So stripping reactions exhibit mainly forward scattering and have large cross sections due to the high b values for which reaction is probable. Several models have been proposed to account for their quantitative behaviour.

The spectator model

This assumes that in the reaction $A + BC \rightarrow AB + C$, the fragment C acts as a 'disinterested spectator' with the same velocity before and after reaction. Thus the motion of C relative to the centre of mass of $B..C$ is ignored. This model predicts that products are scattered exactly forward and that there is an unique small product translational energy. The model is successful in the broadest qualitative terms but fails quantitatively.

Electron jump mechanism (harpoon model)

Polanyi's early sodium flame experiments[7] indicated that sodium–halogen reactions had very large cross sections and Magee[8] proposed an electron transfer model. This can be represented in the alkali atom–halogen molecule case as:

$$M + X—X_1 \rightarrow M^+ + X^-—X_1 \rightarrow M^+—X^- + X_1$$

An approximate treatment assumes that initially the $X..X_1$ distance remains near its most probable value and the discussion focuses on the transfer of the electron between M and atom X.

In this essentially two-body picture the energy of ionic and covalent configurations for $M..X$ may be schematically represented as in figure 9.14, where the 'non-interacting' X_1 is shown in brackets. M approaches X (r decreasing) on the covalent surface which eventually crosses the ionic surface at r_c. At this distance the electron transfer takes place, $M...X(X_1) \rightarrow M^+..X^-(X_1)$ and the non-interacting X_1 separates from the now ionic molecule $M^+\ X^-$. This crossing of surfaces is a forbidden process and it is more correct to state that within a small range of $M..X(X_1)$ distances the charge distribution of the lowest state abruptly re-arranges to $M^+..X^-(X_1)$, with the force between the charged species becoming strongly attractive. This has been phrased more

[7] M. Polanyi, *Atomic Reactions*, Williams and Norgate, 1932.
[8] J. L. Magee, *J. Chem. Phys.*, **8**, 687 (1940).

dramatically: 'atom M tosses out its valence electron to hook atom X which it then hauls in with its Coulombic attraction'[9]. The critical distance r_c should be calculable from the energy required to form the ion pair*:

$$\left(\frac{1}{4\pi\varepsilon_0}\right)\frac{e^2}{r_c} = I - A$$

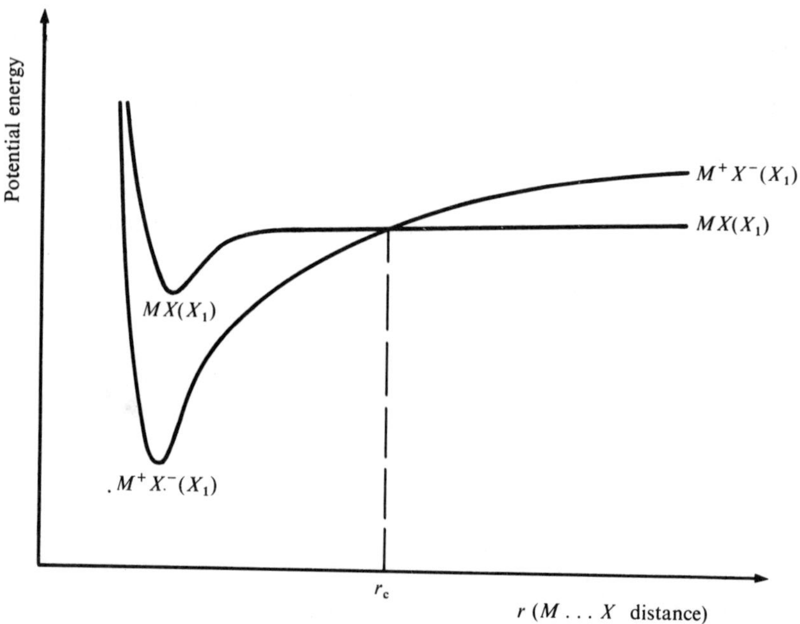

Figure 9.14
Potential energy diagrams for the ionic and covalent, approximately two-body case—$M^+X^-(X_1)$ and $MX(X_1)$ respectively. The curves cross at $r = r_c$.

where I is the alkali metal ionization energy and A the 'vertical' electron affinity of the $X\,X_1$ molecule†. The reaction cross section is given by $S(E) = \pi r_c^{\,2}$.

Again this simple model has some qualitative merit but it cannot be modified to give a wide range of quantitative success. It is not clear what features must be incorporated into the relevant potential energy surfaces to reproduce the experimental characteristics

[9] D. R. Herschbach, *Adv. Chem. Phys.*, **10**, 319 (1966).

* The constant term $1/4\pi\varepsilon_0$ arises when electrostatic energy is expressed in SI units. ε_0 is 'the permittivity of a vacuum' and has the value $8{\cdot}854 \times 10^{-12}$ F m^{-1}. Thus if the electronic charge is given in coulombs and r in metres the energy is in joules.

† The 'vertical' electron affinity is for formation of X^-—X_1 at the same internuclear distance as that in X—X_1. This quantity is generally smaller than the electron affinity of atomic X which can be used in the simplest two-body model.

of stripping reactions, though some form of electron transfer model should provide the basis of a satisfactory approach.

9.7 Long-Lived Complexes

Some reactions show isotropic product distributions, symmetrical about centre of mass scattering angle $\theta = 90°$. This is illustrated in figure 9.15 which shows the minimum intensity at $\theta = 90°$, and intensity increasing to symmetrical maxima at $\theta = 0°$ and 180°. This should be compared with figure 9.12 which demonstrates anisotropic scattering. Examples of such symmetrical distributions are found in the reaction between: alkali atoms and alkali halides (e.g. $Cs + RbCl \rightarrow CsCl + Rb$); alkali atoms and polyhalide molecules; chlorine or fluorine atoms and olefins; oxygen atoms and halogen molecules.

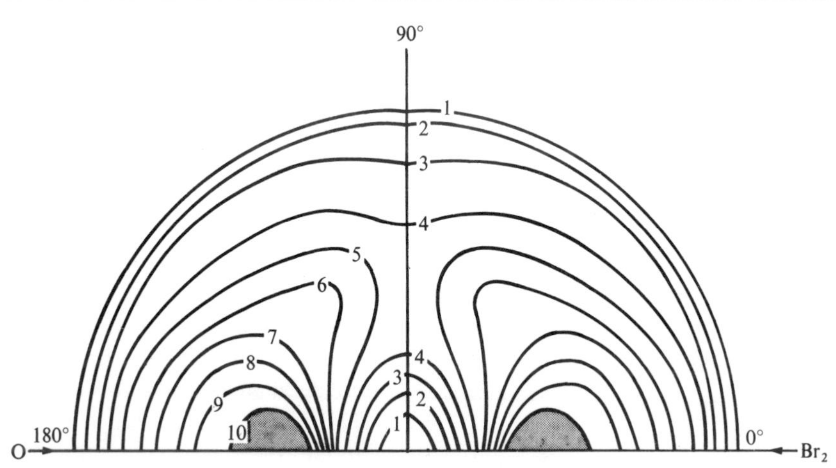

Figure 9.15
The reaction $O + Br_2 \rightarrow BrO + Br$.

Part of the BrO product angular and velocity distribution diagram at one relative reactant energy. The maximum probability is highlighted by shading and shows the forward and backward symmetry characteristic of reactions proceeding via a long-lived collision complex—compare figure 9.12.

These product distributions indicate a collision complex undergoing several rotations and hence many vibrations. By the time a complex has completed a rotation it has 'forgotten' the direction at which the reactants approached. When complexes exist for a few rotations, on dissociation the particles fly apart with equal probability in all directions, giving isotropic scattering. Why then is the product intensity maximum at 0° and 180° and minimum at 90° *in the experimental plane*? This arises because for each scattering angle θ there is a cone of scattering at this angle about the direction of the initial relative velocity vector. Such a cone is illustrated in figure 9.8, though for elastic scattering in that example. Since each θ has the same probability, the total intensity of products within the cone is the same for each θ. However at scattering angles of 0° and 180° to the initial relative velocity, the entire cones are concentrated along the initial relative velocity axis giving very high product intensities at a detector in the experimental plane. For scattering at 90°, the cone spreads out into a disc about the initial relative velocity axis and only a very small fraction of product intensity lies in the experimental plane where the detector is placed. So the angular distribution in the experimental plane has two intensity maxima at 0°, 180°, symmetrical about the minimum at $\theta = 90°$, as shown in figure 9.15. Some intermediate cases have been observed where the lifetime of the complex appears to be around one rotational period. They are somewhat whimsically referred to as 'osculating' complexes. Collision complexes should be associated with a shallow basin in the potential energy surface and they must not be confused with that purely hypothetical entity, the activated complex in activated complex theory.

A modified RRKM theory can be applied to the unimolecular dissociation of a complex with a lifetime of several rotational periods[10], and the experimental angular and velocity distributions for products have been successfully explained.

9.8 Energy Distribution in Products

This is one of the few aspects of detailed collision studies where more sophisticated information has been provided by another technique—chemiluminescence studies as described in the next chapter.

Several examples of determining energy partitioning in products by beam experiments have already been quoted. The most important trends are high internal energy for products of stripping reactions, and relatively low internal excitation with products of rebound reactions. An ingenious example of a more detailed determination of product excitation is for the reaction $F + D_2 \rightarrow DF + D$.[11] In this case the separation of vibrational levels in product DF is large, and should affect the velocity and angular distribution of products. Several peaks were observed in the experimental angular distribution and they were correlated with different scattering patterns for each DF vibrational state. The most prominent were for the $v = 3$ and $v = 4$ product vibrational states. This result is mirrored by quantum mechanical calculations which show that distribution of vibrational energy is sensitive to reactant collision energy.

9.9 Steric Effects

Conventional chemical kinetics indicated long ago that reactions have preferred orientations for the colliding reactant species, but rate constant data are not easy to interpret in detail and the theoretical descriptions of rate constants do not give any reliable predictions in this respect. In molecular beam experiments with polar reactants, electrical fields can be used to impose a predominant orientation on colliding

[10] Discussed by D. R. Herschbach, *Disc. Faraday Soc.*, **55**, 247 (1973).
[11] T. P. Schater *et al.*, *J. Chem. Phys.*, **53**, 3385 (1970).

reactants and this can be reversed by reversing the fields. It has already been seen that the $K + CH_3I$ reaction is favoured when K approaches the I end of methyl iodide. In the analogous $Rb + CH_3I$ reaction a better definition of methyl iodide orientation was obtained[12]: (a) by velocity selection; and (b) the orientation field assembly was tilted to be essentially parallel to the most probable relative velocity vector. Approach of Rb to the I end of methyl iodide was three times more efficient for reaction than approach to the CH_3 end. The story with the $K + CF_3I$ reaction is rather different[13]. When K approaches the I end of the molecule the KI product is scattered backwards as in the CH_3I rebound reaction. However, when K approaches the CF_3 end there is still a high probability of reaction to KI, but this time with forward scattering. It seems that K flies past the CF_3 group to strike the iodine atom which is not as well shielded by CF_3 as it is by CH_3.

9.10 Endothermic Reactions: the Effect of Reactant Internal Energy

Most reactions studied in beams have large cross sections at low energy in order to give sufficient product intensity under beam conditions, so these reactions have small threshold energies and are usually highly exothermic. The reactants are in their ground electronic and vibrational states and their translational energies are small compared with the reaction's exothermicity, so to a good approximation such reactions have been studied at one total energy for the system.

Endothermic reactions have appreciable threshold energies and sizable activation energies in the bulk kinetic systems. Information about them has been limited to applying the principle of microscopic reversibility to the exothermic reactions studied in beams[14]. However, as stated above, this is equivalent to a single total energy for the reacting system. The results for the exothermic reaction usually show a most probable product energy distribution for the given total energy, E:

$$A + BC \rightarrow AB^{\dagger} + C^{*} \qquad \text{at } E \tag{9.10.I}$$

Thus, for the reverse reaction, this tells us that at total reactant energy E the energy distribution corresponding to the highest probability of reaction is the converse of (9.10.I):

$$\text{At } E \qquad AB^{\dagger} + C^{*} \rightarrow A + BC \tag{9.10.II}$$

This result has no bearing on the macroscopic activation energy for reaction (9.10.II), as this relates to bulk gases with equilibrium energy distribution for reactants.

It is usual to consider that activation energy is governed largely by translational energy, perhaps due to the relative ease with which appropriate theoretical models can be constructed. This assumption holds best over the temperature range where concentration of excited vibrational states is negligible. However, the non-statistical product distribution found by beam experiments and chemiluminescence studies for exothermic reactions shows that internal energy must be taken into account in endothermic reactions.

[12] R. B. Bernstein and R. J. Beuhler Jr., *J. Chem. Phys.*, **51**, 5305 (1969).
[13] P. R. Brooks, *Disc. Faraday Soc.*, **55**, 299 (1973).
[14] Discussed in J. L. Kinsey, *J. Chem. Phys.*, **54**, 1206 (1971).

Recently, however, it has become possible to investigate experimentally the effect of translational and internal energy in endothermic reactions and at present this subject is being pursued with great vigour. Some early results are:

(a) The reaction $HI + DI \rightarrow HD + I_2$ has an activation energy of 185 kJ mol^{-1}. A molecular beam investigation[15] used a single, HI-seeded, supersonic beam entering a chamber containing a low pressure of DI. Analysis for HD product was carried out by mass spectrometry. With reactant relative translational energy in the range 80–450 kJ mol^{-1}, going well above the macroscopic activation energy, no reaction was observed. Thus under thermal conditions, internal excitation, which is largely absent from the beam experiment, must play a significant role.

(b) Early beam studies of the reaction K + HCl, where HCl was in its ground vibrational state, indicated a very low probability of reaction with a cross section of about 0·0015 nm^2. A beam experiment has now been conducted[16] with HCl in the first excited vibrational level produced by laser irradiation. The $v = 1$ level is 33 kJ mol^{-1} above the ground state and this gave a dramatic enhancement to reactivity, the reaction cross section increasing a hundredfold. Unfortunately an experiment with translational energy as high as 33 kJ mol^{-1} has not been carried out for comparison.

Further Reading

D. R. Herschbach, *Adv. Chem. Phys.*, **10**, 319 (1966).

J. L. Kinsey and J. I. Steinfeld, *Prog. Reaction Kinetics*, **5**, 1 (1970).

J. L. Kinsey in *MTP Int. Rev. Sci., Phys. Chem. Ser. 1*, Vol. 9, (Ed. J. C. Polanyi), Butterworths, 1972.

D. R. Herschbach, *Disc. Faraday Soc.*, **55**, 233 (1973).

R. D. Levine and R. B. Bernstein, *Molecular Reaction Dynamics*, Oxford University Press, 1974.

Exercises

9.1
In a molecular beam study of the reaction $M + RI \rightarrow MI + R$, small-angle scattering showed that the RI beam, which was 3 mm thick with a concentration of 10^{12} molecules cm^{-3}, produced a 5% loss of intensity from the M beam.

Two detectors, one for $MI + M$ and one for M alone, were used to study product angular distribution and they revealed that of all scattered M, 4·0% formed MI. Evaluate the reaction cross section.

(Ans: 0·28 nm^2)

9.2
In reactions between an alkali metal atom and (a) a halogen (b) an alkyl halide, at the same relative translational energy for reactants, the variation of differential cross section with centre of mass scattering angle is shown in the table:

[15] J. B. Anderson and S. B. Jaffee, *J. Chem. Phys.*, **51**, 1057 (1969).
[16] T. J. Odiorne, P. R. Brooks and J. V. V. Kasper, *J. Chem. Phys.*, **55**, 1980 (1971).

Scattering angle in degrees		0	15	30	45	60	75	90	105	120	135	150	165	180(π)
$\left(\frac{\text{diff. cross section}}{10^{-2}\ \text{nm}^2\ \text{sr}^{-1}}\right)$	(a)	43	42	40	30	24	18	15	12	10	9	10	10	11
	(b)	0	0	0·1	0·6	1·4	2·2	3·0	3.8	4·4	5·0	5·3	5·5	5·5

Evaluate graphically the reaction cross section for (a) and (b). To which class of reaction does each belong? When performing the final integration remember to express the scattering angle in radians (0°–180° becomes 0–π radians).

(Ans: (a) 2·1 nm^2; (b) 0·30 nm^2)

9.3

In the reaction $M + X_2 \rightarrow MX + X$ where the atomic weights of M and X are 40×10^{-3} and 80×10^{-3} kg respectively, a molecular beam study showed that the most probable laboratory angle for the scattering of MX was 0°, measured from the direction of the M beam. The selected velocities for the M and X_2 beams were 3×10^2 and 4×10^2 m s^{-1} respectively. The most probable relative translational energy corresponding to the scattering of MX at 0° was 32 kJ mol^{-1}.

Draw the Newton diagram for reactants and the most probable product scattering. Compare the result with those given in figures 9.10, 9.11 and 9.13.

9.4

In a beam experiment where the beams cross at right angles, one reactant, an atom of atomic weight 50×10^{-3} kg, has a selected velocity of 8×10^2 m s^{-1}. The other reactant, molecular weight 150×10^{-3} kg, is not velocity selected and its most probable velocity is 2×10^2 m s^{-1}. At the source temperatures the reactants are in their ground electronic states and the molecular reactant is in the ground vibrational state with an average rotational energy of 2·2 kJ mol^{-1}.

The angular and velocity distribution of products shows that this is a rebound reaction with most probable product relative translational energy of 50 kJ mol^{-1}. Given that the reaction is 85 kJ mol^{-1} exothermic, calculate the fraction of available energy that appears as internal energy of products.

This reaction proceeds by a direct or impulsive mechanism and the duration of the collision is of the same order of magnitude as the time taken for molecules to traverse a typical distance over which molecular interaction is significant, say 1·0 nm. What is the order of magnitude of the duration of collision?

(Ans: 50% of available energy; duration $\sim 10^{-12}$ s)

9.5

A simple model for stripping reactions is given in section 9.6. The ionization potentials of K, Cs and Rb are 428, 402 and 382 kJ mol^{-1} respectively. The vertical electron affinity of I_2 has been estimated as 154 kJ mol^{-1}. Calculate the reaction cross sections for alkali metal + iodine reactions on this model and compare the results with the experimental values given in the table:

Reaction	$K + I_2$	$Rb + I_2$	$Cs + I_2$
Reaction cross section/nm^2	1·27	$\gtrsim$1·50	1·95

Does the comparison support the conclusions drawn in section 9.6?
(For the calculation in SI units use $\varepsilon_0 = 8{\cdot}85 \times 10^{-12}$ F m^{-1})

Chapter 10: Chemiluminescence, Hot-Atom Reactions, Ion–Molecule Reactions

(A) Chemiluminescence

Chemiluminescence is the emission of radiation from the products of an exothermic reaction when these products are formed in a state of thermal disequilibrium with respect to their surroundings. The excited states with higher than equilibrium concentrations may be electronic, vibrational and rotational, and since their distributions cannot be described by characteristic temperatures the population of each state must be specified. Observation of chemiluminescence thus provides information on the way reaction exothermicity is channelled into different energy modes in the products and this furnishes some insight into the shape of the potential energy surface for the reaction.

Where the reaction releases enough energy for products to be formed in electronically excited states the possibility of crossing to another potential energy surface arises. This topic is treated at length in a recent review[1], but the present section will concentrate on the important cases where exothermicity appears as vibrational, rotational or translational excitation.

10.1 Experimental Methods[2]

The aim of these investigations is to measure the initial excitation of the newly formed product, but the experimenter observes the excitation after some distortion by

[1] T. Carrington and J. C. Polanyi, *MTP Int. Rev. Sci., Phys. Chem. Ser. 1*, Vol. 9, (Ed. J. C. Polanyi), Butterworths, 1972.

[2] Reviewed by T. Carrington and D. Garvin in *Comprehensive Chemical Kinetics*, Vol. 3, (Eds C. H. Bamford and C. F. H. Tipper), Elsevier, 1969.

relaxation processes that occur between formation and observation. For gas phase studies relaxation takes place either by emission of radiation or on collision. The relative rates for these alternatives are such that at S.T.P. collisional de-activation predominates for the excited vibrational states but at sufficiently low pressures (~ 1–10^{-2} Torr, ~ 133–$1{\cdot}3$ Nm^{-2}) the influence of collisional processes becomes negligible and emission can be detected in the infrared region (see sections 2.10, 11). In a typical system:

$$A + BC \rightarrow AB + C$$

reaction results in AB being formed in vibrational and rotationally excited states and it is possible, with modern techniques, to work at sufficiently low pressures for easy detection of the infrared chemiluminescence due to vibrational transitions. On this basic method there are several variants which differ in the way they attempt to extract information concerning the initial product energy distribution from the partially relaxed distributions that are observed.

(i) Fast flow with single observation[3]. After mixing, the reactants pass a single observation point placed as close as possible to the mixing point. The conditions are chosen to minimize the relaxation that has occurred before the infrared emission is recorded.

(ii) Flow with measured relaxation. Here several windows are placed at intervals along the line of flow. These observation points correspond to a series of times after mixing reactants and so they record progressive relaxation with time. A graphical extrapolation back to the time at which reaction occurred gives quite dependable values for the relative probability of forming various vibrational states. The method has been applied in great detail[4] to the classic example in this sphere, $H + Cl_2 \rightarrow HCl^\dagger + Cl$.

(iii) Arrested relaxation. The chemiluminescent reaction takes place when two uncollimated beams meet in the centre of a reaction chamber at very low background pressure. Products are removed rapidly, either at a cold wall or by fast pumping, so they do not remain in the reaction/observation zone long enough to undergo serious relaxation. Thus the chemiluminescence reflects a steady-state distribution very close to the initial distribution on reaction[5]. This has been confirmed in some cases by the detection of unrelaxed rotational excitation—the de-activation of rotationally excited states in collision is much more efficient than that of vibrational states, so vibrational relaxation by collision is certainly negligible in such studies. Where rotational and vibrational excitation is measured, the translational excitation can be calculated accurately from the known heat of reaction.

(iv) Chemical lasers[6]. Lasing action can be obtained from a chemical reaction if population inversion (excess of excited state over ground state population) in products is achieved with sufficient rapidity. Then laser gain can exceed the threshold value and stimulated emission compete successfully with the relaxation processes. The most

[3] e.g. J. K. Cashion and J. C. Polanyi, *Proc. R. Soc.* (*London*), **A258**, 529, 564, 570 (1960).
[4] e.g. P. D. Pacey and J. C. Polanyi, *J. Appl. Opt.*, **10**, 1738 (1971).
[5] e.g. H. Heydtmann and J. C. Polanyi, *J. Appl. Opt.*, **10**, 1738 (1971).
[6] e.g. M. J. Berry and G. C. Pimental, *J. Chem. Phys.*, **49**, 5190 (1968); *J. Chem. Phys.*, **53**, 3453 (1970).

common way of initiating the reaction is with a flash of light. A well known example is the use of CF_3I as a source of fluorine atoms:

$$CF_3I + h\nu \rightarrow F^{\cdot} + CF_2I^{\cdot}$$

and in the presence of hydrogen this is followed by the reaction to be investigated:

$$F^{\cdot} + H_2 \rightarrow H^{\cdot} + HF^{\dagger}$$

The vibrationally excited HF emits a quantum of infrared radiation

$$HF^{\dagger} \rightarrow HF + h\nu'$$

and the photon can stimulate emission from another molecule in precisely the same energy state (*unrelaxed* excited HF)

$$HF^{\dagger} + h\nu' \rightarrow HF + 2h\nu'$$

The two photons can produce further stimulated emission and so on to give a cascade of photons and lasing action. The actual laser gain depends on the extent of population inversion and by measuring the gain for different transitions under different experimental conditions it is possible to estimate the relative populations in the initial vibrational distributions of product molecules.

10.2 Results: Implications for Potential Energy Surfaces and Energy Release

The most accurately measured energy distributions are for reactions of the type $A + BC \rightarrow AB + C$ and some examples are given in table 10.1. Absolute yields cannot be measured by these techniques which give only the relative rates at which vibrational and rotational states are formed. Absolute values of total rate coefficients can be taken from conventional kinetic studies, and chemiluminescence data used to assign the relative rates to the various product states.

Table 10.1
The fraction of reaction exothermicity appearing as vibrational, rotational and translational energy, and the vibrational state of highest population. Data are extracted from T. Carrington and J. C. Polanyi, *MTP Int. Rev. Sci., Phys. Chem., Ser. 1*, Vol. 9, (Ed. J. C. Polanyi), Butterworths, 1972.

Reaction	Exothermicity/ kJ mol^{-1}	Vibrational state of highest population	Mean fraction of energy appearing as vibration (V) rotation (R) or translation (T)		
			f_V	f_R	f_T
$H + Cl_2 \rightarrow HCl + Cl$	202	2	0·39	0·07	0·54
$D + Cl_2 \rightarrow DCl + Cl$	208	3	0·40	0·10	0·50
$H + Br_2 \rightarrow HBr + Br$	183	4	0·56	0·05	0·39
$F + H_2 \rightarrow HF + H$	145	2	0·67	0·07	0·26
$F + D_2 \rightarrow DF + D$	144	3	0·69	0·06	0·25
$Cl + HI \rightarrow HCl + I$	141	3	0·71	0·13	0·16
$Cl + DI \rightarrow DCl + I$	142	5	0·71	0·14	0·15

Potential energy surfaces for exothermic reactions were discussed in section 5.4 which introduced the useful concept of attractive and repulsive types, and mixed energy release. Attractive surfaces have the downhill slope from the energy barrier located along the direction of approach of reactants. Attractive release and some mixed energy release are expected to result in exothermicity appearing largely as vibrational energy of products. Repulsive surfaces have the downhill slope located along the direction of separation of products. Repulsive and mixed energy release should yield a lower fraction of exothermicity as vibrational excitation, more as translational. If the reactant molecule contains heavy atoms and the approaching atom is light, mixed energy release is diminished and an anomalously high fraction of exothermicity can appear as translational excitation.

Another way of expressing the relative position of the energy barrier on attractive and repulsive surfaces is that barriers displaced into the reactant or entrance valley are early barriers (attractive case) whereas barriers displaced toward the product or exit valleys are late barriers. Chemiluminescence data, as exemplified in table 10.1, and complementary molecular dynamic trajectory calculations are providing progressively more detailed pictures of the nature of these surfaces.

(i) For comparable reactions, such as related families of exchange reactions, there tends to be a strong correlation between earlier barrier location (with increased energy release into vibration) and decreasing barrier height. This is illustrated by comparing the $H + Br_2$ reaction with the $H + Cl_2$ where the former has a lower energy barrier and the HBr product has higher vibrational excitation than HCl.
(ii) The conversion of available energy into vibration tends to be lower for H atoms than for heavier reactant atoms such as halogens, which can be interpreted as evidence of the light atom anomaly.

(iii) In exothermic four-atom exchange reactions:

$$A + BCD \rightarrow AB^{\dagger} + CD$$

trajectory calculations indicate that the 'new' bond takes up most of the exothermicity as vibrational energy. This is not surprising since the energy is released on formation of $A\,..\,B$, whereas the existing CD bond is not subjected to such strong perturbation. The predominant excitation of the new bond was first observed in flash photolysis studies for reactions such as $O + NO_2 \rightarrow O_2{}^{\dagger} + NO$; $O + ClO_2 \rightarrow O_2{}^{\dagger} + ClO$ (see section 4.1) and it has been found in chemiluminescent studies[7] for the reaction $H + NOCl \rightarrow HCl^{\dagger} + NO$. Very high vibrational excitation also occurs[8] for the reaction $H + O_3 \rightarrow OH^{\dagger} + O_2$, so there is no evidence in such cases for the light atom anomaly on repulsive surfaces, and the surface is considered to be attractive.

(iv) For highly exothermic reactions with low and early energy barriers the low energy required to surmount the barrier can probably be provided by translational energy alone. In the endothermic direction, the reverse reaction, there is a high, late barrier and the degree of freedom most conducive to reaction may be vibration in reactants. Evidence on this point is provided by molecular beam experiments discussed in section 9.10. The relative importance of translational and internal energy in overcoming the potential energy barrier is the subject of intense practical and theoretical investigation at present.

[7] P. E. Charters, B. N. Khane and J. C. Polanyi, *Disc. Faraday Soc.*, **33**, 107 (1962).
[8] P. E. Charters, R. G. Macdonald and J. C. Polanyi, *J. Appl. Opt.*, **10**, 1747 (1971).

(B) Hot-Atom Reactions

Molecular beam experiments have been fruitful in probing exothermic reactions and some endothermic reactions but although collision energies up to about 400 kJ mol^{-1} have been investigated, most work has been carried out at much lower energies. Thus the reactions studied have been those which would also be observed in conventional thermal systems, where the reactants are at approximate Maxwell–Boltzmann equilibrium energy distribution. However, other reaction modes are possible at higher collision energies—for the same reactants there are other reactions with higher threshold energies. The familiar reaction modes are merely those with the lowest threshold energy in that system, usually called *thermal reactions* since they are observed in conventional thermal systems. Reaction modes with higher threshold energies may be referred to as *hot reactions*. These possibilities are illustrated for a hypothetical general case in figure 10.1. For some years techniques have been available for investigating these high energy reactions. They rely mainly on nuclear recoil methods for producing atoms of extremely high translational energy and for this reason the field is usually known as hot-atom reactions. With nuclear recoil sources, atoms are produced at energies far above those at which chemical reactions occur, but the atoms lose energy by collision and eventually reach the chemical reaction energy range (0–1000 kJ mol^{-1}). Within this range the hot atoms have all possible energies, so there is *no* selection of translational energy, though the addition of inert gases to increase the probability of collisional energy loss may lower the average reactant energy. Perhaps the main importance of hot-atom studies is in making it possible to look into the new kinetic world of high energy reaction modes.

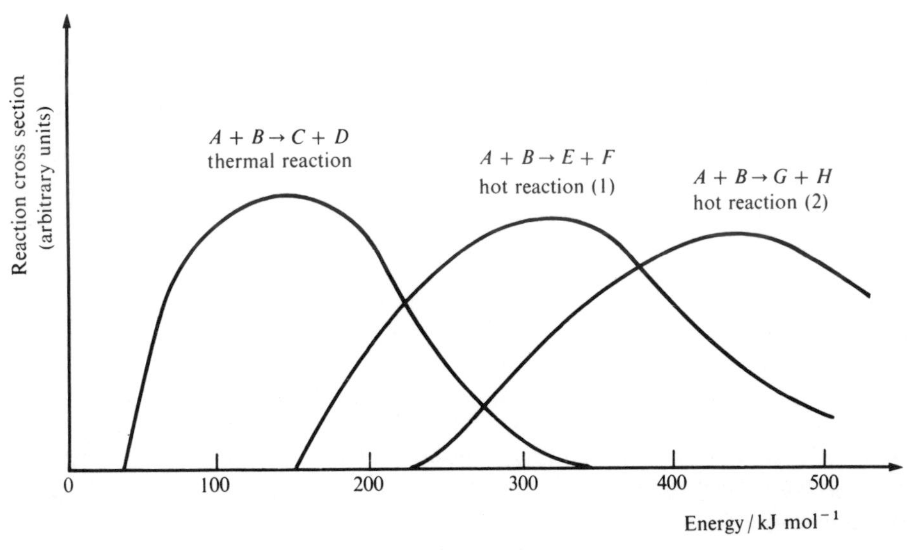

Figure 10.1
Excitation functions for the thermal and two hot reactions in a hypothetical case.

10.3 Hot Atoms from Nuclear Sources

The most important type of hot-atom source[9] is in nuclear reactions where atoms are formed with initial energies in the range 10^7–10^8 kJ mol^{-1}. At such energies the species exist as polyvalent ions but they are neutralized before reaching the chemical reaction energy range. Typical nuclear reactions are:

$$^{3}He + n \rightarrow p + {}^{3}H(\text{tritium, T})$$

$$^{12}C + \gamma \rightarrow n + {}^{11}C$$

$$^{19}F + \gamma \rightarrow n + {}^{18}F$$

Thus the high energy atoms are formed as radioactive isotopes. This does not mean, of course, different chemical behaviour from the normal isotope, but products containing the atom are radioactively labelled and can be detected in very small concentrations by tracer techniques. End-product analysis is carried out for very low product yields using g.l.c. with detection of the radioactive species by standard radiation-counting methods.

It is important to distinguish between products formed by reaction of atoms which still possess high energy, and products from atoms which have lost the high translational energy in collisions and react at thermal energies. The most useful criteria for deciding between hot or thermal reactions are:

(i) Observing a product that is not found in the conventional thermal system.

(ii) Use of a scavenger, ideally one with very low reactivity for hot species and high reactivity for thermal atoms. The hot product yield should be independent of scavenger concentration while thermal yields decrease rapidly as scavenger concentration is increased.

(iii) Unlike thermal reactions, hot reactions are insensitive to temperature.

(iv) Addition of inert gas, which increases the relative probability of collisional energy loss and so progressively suppresses hot reaction. In this context such inert gases are referred to as *moderators*. Lower energy conditions can also be attained by studying reactions in liquid or solid phases or, in the extreme case, by suspending the reactant in a solid inert gas matrix.

Although, in principle, hot-atom chemistry could be carried out for any element, very little is known about even thermal reactions for the atoms of most elements, and hot-atom studies are limited to half a dozen species. By far the most intensively studied atom is the hydrogen isotope, tritium (T), and there is some detailed work with halogens, particularly ^{18}F. Reactions of such monovalent atoms often occur in only one reactive step and their chemistry is straightforward compared with that of polyvalent species. The best understood in the latter category is ^{11}C whose properties will be outlined following a brief account of T reactions.

[9] For a review of experimental methods see R. L. Wolfgang, *Prog. Reaction Kinetics*, **3**, 97 (1965).

10.4 Hot-Atom Reactions of Tritium

The hot-atom reactions of tritium with a wide variety of hydrocarbons and substituted hydrocarbons are of four basic types[10].

(i) *Abstraction* of hydrogen atoms

$$T^* + RH \rightarrow HT + R^{\cdot}$$

This is also the thermal reaction for hydrogen atoms.

(ii) *Substitution*

$$T^* + RX \rightarrow RT + X^{\cdot}$$

(iii) '*Double*' *displacement*

$$T^* + RCH_3 \rightarrow RCHT^{\cdot} + 2H^{\cdot}$$

(iv) *Addition* to a π bond system

$$T^* + \rangle C{=}C\langle \rightarrow (\rangle CT{-}C\langle\, .)$$

In reactions of types (ii)–(iv) the products may be highly excited and undergo secondary isomerization or decomposition. Such secondary products can usually be distinguished from primary products by techniques which diminish reactant energy or increase the possibility of de-activating the excited primary products and this is usually achieved simply by varying pressure or adding moderators.

The experimental results do not give direct information on how the excess energy is disposed during the course of reaction. The variation in yield of primary and secondary products with reactant pressure suggests a broad distribution of excitation energy and that a substantial fraction of product molecules are capable of secondary isomerization or decomposition. Experiments in liquids and solids indicate that in a small fraction of reactions there are very high excitation energies. In more detail, the way results vary with different reactant molecules suggests that the residual excitation energy is a function of the chemical environment of the individual C—H bond involved in the substitution. The nature of the specific bond is also a main factor in abstraction reactions, where the yield for abstraction of H atoms from C—H bonds shows very good correlation with the bond dissociation energies.

For substitution reactions, if the reactant molecule contains an asymmetric carbon atom the stereochemistry of the process becomes of prime interest. In the gas phase, substitution with retention of configuration is the predominant mode, e.g. with $(CHFCl)_2$. These experiments have been carried out only for reactant molecules containing some heavy atoms or groups and do not give direct information about the stereochemistry of substitution where only light atoms are involved—as in CH_4, for example. Recent molecular dynamical calculations indicate that for the latter case inversion is important only at low reactant energies[11].

[10] For a recent general review see F. S. Rowland, *MTP Int. Rev. Sci., Phys. Chem., Ser. 1*, Vol. 9, (Ed. J. C. Polanyi), Butterworths, 1972.

[11] Discussed in D. L. Bunker, *Accts Chem. Res.*, **1**, 195 (1974).

10.5 Reactions of Carbon Atoms

With hydrogen and halogen atoms the thermal reactions are well known, and this is very helpful in interpreting hot-atom results. With polyvalent atoms such as carbon there is no information on their thermal chemistry and this behaviour must be deduced from data on moderated hot atoms[12]. Most attention has been given to hydrocarbon systems and it has been established that the reactions of hot and thermal carbon atoms are qualitatively similar due to the high intrinsic reactivity of the atom.

With hydrogen atoms the primary product often survives but this is rarely the case with carbon atoms. The main primary processes are:

(i) Insertion into the C—H bond, for example

$$CH_3CH_3 + {}^{11}C \rightarrow CH_3CH_2{}^{11}CH^*$$

$$CH_2{=}CH_2 + {}^{11}C \rightarrow CH_2{=}CH{-}{}^{11}CH^*$$

(ii) Insertion into the C=C bond, for example

$$CH_2{=}CH_2 + {}^{11}C \rightarrow \underset{\diagdown\ {}^{11}C\ \diagup}{CH_2{-}CH_2^*}$$

(iii) H abstraction to give ^{11}CH or $^{11}CH_2$. This reaction mode seems to play only a minor role in the system.

Product yields are governed by secondary processes involving the excited intermediates formed in primary reactions. These are summarized below for ethylene.

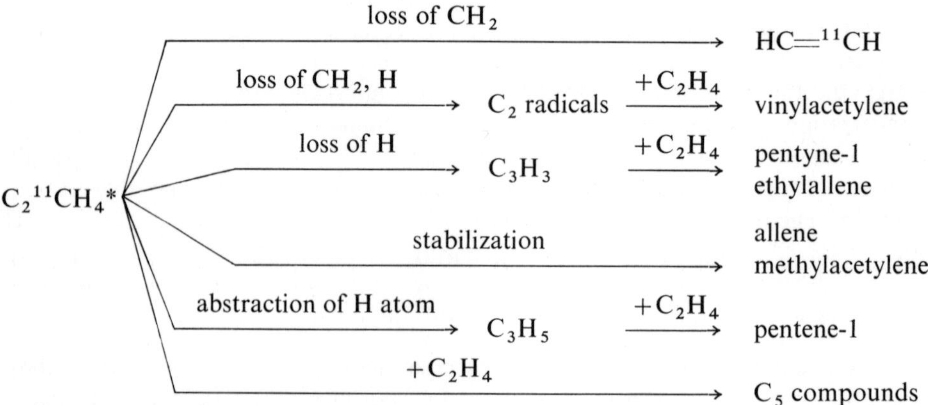

The relative probability of the different reaction paths depends on: (a) the spin state of the carbon atom, which can be in a singlet or triplet state; (b) the kinetic energy of the atom; (c) the efficiency with which excitation energy is removed from the intermediate in molecular collisions.

Moderator studies provide important insights into the different paths and the extreme case, where reactant molecules are suspended in a solid matrix of the moderator, such as krypton, gives the clearest picture of the behaviour of thermal carbon atoms. These

[12] R. L. Wolfgang, *Prog. Reaction Kinetics*, **3**, 97 (1965).

form relatively low energy intermediates in the primary step and the intermediates are rapidly de-excited. However, although the stabilized intermediate (allene) becomes the major product, C_5 adducts and the fragmentation product acetylene are still found in low yield[13].

10.6 Photochemical Methods

If diatomic molecules absorb monochromatic radiation for which the energy of the quanta exceeds that required to dissociate the molecule, the excess energy appears as translational energy of the separating atoms. The translational excitation is lower if one of the atoms is formed in an electronically excited state. A simple calculation based on the conservation of momentum and energy shows that the translational excitation is shared between the two atoms in the inverse ratio of their masses. Thus in the photolysis of HI, for example, because of the great disparity in atomic masses the hydrogen atom retains almost all the available translational energy. Various metal arc discharge lamps combined with filters or monochromators provide intense monochromatic radiation in the ultraviolet, and HCl and HBr can also be used as sources of hydrogen atoms. This means that many initial energies for the atoms are available in the range from about 300 to 30 kJ mol^{-1}. The method can also be used to prepare hot deuterium or tritium atoms from the corresponding halides. As opposed to hot atoms from nuclear sources, the photochemically generated atoms are initially monoenergetic, with their initial energy well within the chemical reaction range. The subsequent collisions are not all reactive and loss of energy in collisions means that reacting atoms may have energies between the initial value and the threshold energy for reaction. The great usefulness of the method is that the threshold energy can be deduced from the minimum initial energy at which products appear.

Most reactions investigated have been of the atom abstraction type:

$$D + C_2H_6 \rightarrow HD + C_2H_5 \quad \text{or} \quad H + C_2D_6 \rightarrow HD + C_2D_5$$

A fuller reaction mechanism in a typical system, such as deuterium atoms from halide DX reacting with hydrocarbon RH, is:

$$DX + h\nu \rightarrow D^* + X \tag{10.6.I}$$

$$D^* + RH \rightarrow HD + R \tag{10.6.II}$$

$$D^* + DX \rightarrow D_2 + X \tag{10.6.III}$$

$$D^* + DX \rightarrow D + DX \tag{10.6.IV}$$

$$D^* + RH \rightarrow D + RH \tag{10.6.V}$$

The asterisk denotes species with energy above the threshold for the abstraction reaction (10.6.II). All deuterium atoms whose energies fall below the threshold as a result of collisional energy loss in steps (10.6.IV) and (10.6.V) will be scavenged by DX to give D_2. Hence processes (10.6.III), (10.6.IV) and (10.6.V) result in formation of D_2 but the abstraction reaction, (10.6.II), is characterized by HD product. The

[13] C. MacKay, J. E. Nicholas and R. L. Wolfgang, *J. Am. Chem. Soc.*, **88**, 1610 (1966).

relative yield of HD and D_2 after photolysis is measured by mass spectrometry, and the way this yield varies with initial energy of the atom leads to an estimate for the threshold energy of (10.6.II).

If the rate parameter for each process, its average probability in the range from initial energy to the threshold, is denoted by S, then inspection of the mechanism shows that the product ratio is given by:

$$\frac{[D_2]}{[HD]} = \frac{S_{III} + S_{IV}}{S_{II}} \frac{[DX]}{[RH]} + \frac{S_V}{S_{II}}$$

At the same initial energy, the product ratios are measured at a series of different reactant ratios and a plot of $[D_2]/[HD]$ against $[DX]/[RH]$ gives a straight line of intercept $I = S_V/S_{II}$. When deuterium atoms collide with RH, the probability of reaction is $S_{II}/(S_{II} + S_V) = 1/(1 + I)$. This probability is not the true reaction cross section, but it must become zero at the threshold energy. Thus a plot of reaction probability against initial energy will give the threshold energy and a typical result is shown in figure 10.2. At low initial energy correction should be made for the influence of the small proportion of atoms at the higher energy 'tail' of the equilibrium distribution at room temperature.

As stated above, the actual distribution of atom energies in the reactive range depends on energy losses in non-reactive collisions. These have been estimated from the effect of inert gas moderators on product yields using an independent evaluation of the potential energy for interaction between hydrogen atom and moderator. With such data it is possible to 'unfold' the measured reaction probability to give an estimate of the reaction cross section and the result is illustrated in figure 10.3.

The abstraction reactions investigated with photochemical hot atoms are the thermal reactions for these systems. It has also proved possible to study a hot substitution reaction[14]:

$$T + CH_4 \rightarrow CH_3T + H$$

for which the threshold was found to be about 150 kJ mol^{-1}.

(C) Ion–Molecule Reactions

Ion–molecule reactions play a crucial part in electrical discharges through gases, in shock tubes at very high temperatures, in flames, and processes induced by high energy radiation. Furthermore, the chemistry of the upper atmosphere depends critically on such reactions so there is a powerful array of reasons underlying the present rapid growth in ion–molecule studies. The existence of ion–molecule reactions has been recognized since the early days of mass spectrometry[15]. In the mass spectrometer primary ions are formed by electron impact on molecules, and subsequent ion–molecule collisions can give secondary ions. The reaction $H_2^+ + H_2 \rightarrow H_3^+ + H$

[14] C. Chou and F. S. Rowland, *J. Chem. Phys.*, **50**, 2763 (1969).

[15] The historical development of the field is described in L. Friedman and B. G. Reuben, *Adv. Chem. Phys.*, **19**, 33 (1971).

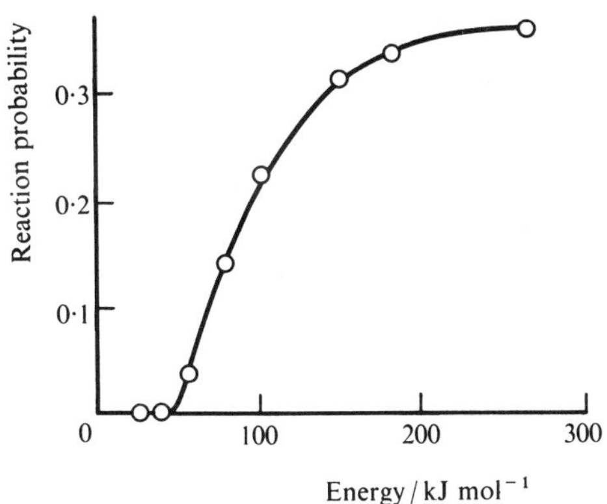

Figure 10.2
Variation of reaction probability with initial atom energy for the reaction $D + C_2H_6 \rightarrow C_2H_5 + HD$ (data from F. Bayrakceken, R. D. Fink and J. E. Nicholas, *J. Chem. Phys.*, **56**, 1008 (1972)).

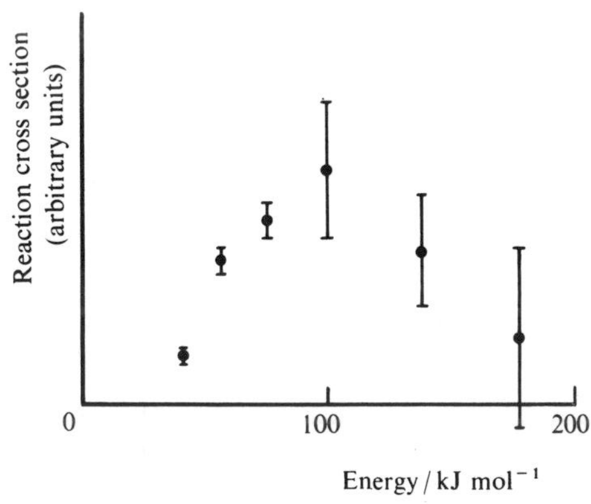

Figure 10.3
Variation of reaction cross section with energy for the reaction $H + n\ C_4D_{10} \rightarrow HD +$ see C_4D_9 (data from R. G. Gann, W. M. Ollison and J. Dubrin, *J. Chem. Phys.*, **54**, 2304 (1971)).

was identified as early as 1925, but until the nineteen fifties most attention was devoted to the primary ionization, to measuring energies at which ions appeared (appearance energy), and to analytical·mass spectrometry. Interest was stimulated by identifying the reaction $CH_4^+ + CH_4 \rightarrow CH_5^+ + H$ since the CH_5^+ ion is a fascinating valency problem for theoretical chemists. About this time improved experimental techniques led to direct investigation of such reactions in the mass spectrometer using the ionization chamber as reaction vessel, and rate constants were measured for various reactions with simple organic molecules.

Rate measurements were progressively extended as experimental sophistication increased and beam methods now push experimental detail towards its limit. Today the growth in such studies is almost exponential. Over the years other methods for forming ions have been added to electron impact. *Chemi-ionization* refers to the production of ions in collisions of electronically excited species. In *photoionization*, molecules absorb quanta of very short wavelength, in the far ultraviolet region. These quanta have sufficient energy to cause ionization and, if monochromatic radiation is used, the identity and initial energy states of the ions may be known precisely. *Field ionization* is the ionization of molecules in the very intense electrical fields that can be produced, for example, by applying a high potential to a very sharp edge. As well as extending the range of analytical mass spectrometry, this allows investigation of reactions such as the unimolecular dissociation of ions.

A brief outline of the major experimental techniques will be followed by a survey of the main features of ion–molecule reactions that emerge from such work, and the chapter concludes with a short account of the chemistry of the upper atmosphere.

10.7 Experimental[16]

A convenient subdivision is into 'swarm' and 'beam' methods. In swarm methods ions are injected into a chamber containing molecules with approximately equilibrium energy distributions. With beam methods either single beams of ions, or beams of both ions and neutral molecules have been used.

Swarms

Reactions may be studied in the ionization chamber of the traditional mass spectrometer where molecules interact with the electrons from the source and are ionized by electron impact. Under a small positive potential from a 'repeller' the ions drift towards the mass spectrometer analyser. At very low pressure these primary ions will reach the analyser without experiencing further collisions, but as pressure increases so does the probability of collision before analysis. If the collision results in a reaction giving secondary ions these will be detected by the analyser in addition to the primary ions. Typically, at about 10^{-2} Torr, $1{\cdot}3$ N m^{-2}, about half the primary ions will collide before reaching the analyser and at about 10^{-1} Torr, 13 N m^{-2}, virtually all will do so.

The time spent by ions in the source region before analysis can be controlled by the instrument's field strength, and rates of ion–molecule reactions can be deduced. The earliest rate constants were measured in this way and the results have proved reliable in many cases.

It is important to remember that, for any parent molecule, mass spectrometry yields more than one type of ion—the decomposition of ions at the electron energies employed in the instrument results in the characteristic mass spectrum of the species. This may complicate the interpretation of ion–molecule processes since product ions could have the same mass/charge ratio as a component of the parent mass spectrum.

[16] Detailed accounts of modern methods are given in E. W. McDaniel *et al.*, *Ion–Molecule Reactions*, Wiley Insterscience, 1970; *Ion–Molecule Reactions*, (Ed. J. L. Franklin), Plenum, 1972; H. S. W. Massey, E. H. S. Burhop and H. B. Gilbody, *Electronic and Ionic Impact Phenomena*, Oxford U.P., 1971.

In addition to this problem the simple method is subject to further drawbacks in that study is limited to a small pressure range and reaction time cannot be controlled independently. The pressure range has been extended with narrower ion source slits and better pumping, and, with other improvements in ion sources, investigations have been extended to pressures where equilibrium constants can be measured. The independent control of reaction times can be achieved by *pulsed ion sources.* Improved discrimination between competing reactions and better rate measurements are obtained with the *flowing afterglow* technique, in which reactions occur in a fast gas flow and a quadrupole mass spectrometer analyses the species present. Where there is a multiplicity of products from one reactant, or where several reactants give the same product a completely different detection technique—*ion cyclotron resonance*—has proved very useful. Positive or negative ions can be 'trapped' for many milliseconds and ion detection sensitivities of lower than 10 ions cm^{-3} achieved. An important refinement is that very well-defined reactant states can be obtained in *photoionization sources* for the ions.

Single-beam methods

The introduction of tandem mass spectrometers was a step forward in isolating specific ion–molecule reactions. The first spectrometer selects the reactant ion and controls the energy of the beam which then passes into a vessel containing reactant molecules. The products are analysed by the second spectrometer which is disposed in one of two ways. The most useful for ion–molecule reactions is where analysis takes place in the same direction as the primary ion beam, but sampling at right angles to this direction is suitable for studying charge transfer or ion decomposition.

Dual beams

As with neutral–neutral reactions, the most detailed picture of ion–molecule reactions comes from dual-beam systems in which reactant states can be well defined. With crossed beams, angular and velocity distributions of products are mapped out, but it is difficult to get absolute product intensities and absolute cross sections in addition to the shape of excitation functions. These, however, can be measured with merging beams.

10.8 Reaction Types

Some of the main features of ion–molecule reactions[17] were established in the first decade of intensive study. In addition to the decomposition reactions which contribute to the range of ions in mass spectra, the main types of ionic reaction are:

(i) charger transfer, e.g. $O_2^+ + NO \rightarrow NO^+ + O_2$

(ii) single-particle transfer $N_2^+ + H_2 \rightarrow N_2H^+ + H$

(iii) condensation $C_2H_4^+ + C_2H_4 \rightarrow C_3H_5^+ + CH_3$

In (iii) the products must result from considerable re-arrangement of a collision complex. A further, rare, category is association e.g.

$$C_3H_7I^+ + C_3H_7I \rightarrow (C_3H_7I)_2^+$$

[17] M. J. Henchman, *Chem. Soc. Ann. Rep.*, **62**, 39 (1965).

Rate constants have been measured for many reactions in the main categories. They tend to be very high—up to 10^{15} mol^{-1} cm^3 s^{-1}. If reactants are regarded as hard spheres then rate constants of this magnitude imply reaction cross sections much greater than the geometric size of the species. This reflects one of the fundamental factors governing these reactions, that the reactants are subject to long-range ion–dipole or ion–induced dipole interaction. For the exothermic processes studied the activation energies are effectively zero, but rate constants show small negative temperature coefficients due to the temperature dependence of collision frequencies.

These points emphasize a fundamental difference between ion–molecule reactions and those of neutral species. Reaction cross sections for neutrals are zero until the threshold energy is reached and then increase to a maximum. Ion–molecule reactions usually have no threshold and cross section decreases with energy, since long-range interactions have less effect as the energy increases. There are always competing processes such as charge transfer on collision or, if the ion is formed at high energy, unimolecular dissociation. The latter process can be very fast and recent field ionization studies have followed the decomposition of excited ions taking place in picoseconds[18], e.g.

$$(CH_3)_3CCH_3^+ \rightarrow (CH_3)_3C^+ + CH_3$$

10.9 Detailed Collision Studies

The use of single beam or crossed and merging dual beams in this field has expanded greatly in recent years. Determining detailed angular and velocity distributions for products has led to important mechanistic conclusions[19], as with molecular beam studies in neutral systems. The main experimental distinction between reaction types is between reactions with anisotropic angular distributions and those with symmetrical scattering about 90° in centre of mass coordinates. The former is characteristic of short-lived direct interaction on collision, and the latter indicates a long-lived collision complex or indirect mechanism. Table 10.2 shows some reactions that have been assigned to these categories.

Table 10.2
Some ion–molecule reactions which have been shown to proceed (a) by direct (b) by indirect mechanism at low collision energies.

(a) Reactions with direct mechanism	(b) Reactions with indirect mechanism
$N_2^+ + H_2 \rightarrow N_2H^+ + H$	$H_2^+ + H_2 \rightarrow H_3^+ + H$
$CO^+ + H_2 \rightarrow COH^+ + H$	$CH_4^+ + CH_4 \rightarrow CH_5^+ + CH_3$
$Ar^+ + H_2 \rightarrow ArH^+ + H$	$C_2H_4^+ + C_2H_4 \rightarrow C_3H_5^+ + CH_3$
$N_2^+ + CH_4 \rightarrow N_2H^+ + CH_3$	

Direct interaction

Product angular distributions are frequently found to be strongly forward peaked, the type associated with stripping reactions. It is also found, for such reactions as

[18] P. J. Derrick and A. J. B. Robertson, *Proc. R. Soc. (London)*, **A324**, 491 (1971); H. D. Beckey, *Field Ionization Mass Spectrometry*, Pergamon, Oxford, 1971.
[19] For a recent review see J. Dubrin and M. J. Henchman in *MTP Inst. Rev. Sci., Phys. Chem., Ser. 1*, Vol. 9, (Ed. J. C. Polanyi), Butterworths, 1972.

Ar^+, N_2^+ or CO^+ with H_2, that the most probable product speeds are very close to those for simple spectator stripping. The cross sections are large and the duration of collisions is very short ($\sim 10^{-14}$ s). Ion–molecule reactions of this simple atom transfer type are common and occur quite efficiently up to surprisingly high energies, well in excess of bond dissociation energies. This must reflect a high probability of transferring an atom in collisions that are just grazing, since a fuller impact must result in some fragmentation.

Potential energy surfaces for ion–molecule reactions often (perhaps always) show basins, though these are very shallow for reactions proceeding by direct interaction at low energy. Until recently it was thought that at *very low* reactant energies all ion–molecule reactions proceed via a long-lived complex, but this has been disproved for reactions of N_2^+ or Ar^+ with D_2 at energies down to 10 kJ mol^{-1}.

Indirect interaction

A substantial number of ionic reactions proceeding via a long-lived complex have now been identified. In all cases, at higher energies, there is a gradual transition from indirect to direct interaction, e.g. for

$$C_2H_4 + C_2H_4^+ \rightarrow (C_4H_8)^+ \rightarrow CH_3 + C_3H_5^+$$
$$C_3H_5^+ \rightarrow C_3H_3^+ + H_2$$

both ionic products show symmetrical angular scattering below about 30 kJ mol^{-1} reactant energy, but as energy increases there is a gradual transition to forward scattering.

The average lifetime of collision complexes in these reactions depends on the initial collision energy and these systems provide valuable tests for theories of unimolecular decomposition.

Role of reactant energy

With endothermic reactions a fundamental question is the relative effectiveness of translational or internal energy in crossing the energy barrier. For the reaction

$$H_2^+ + He \rightarrow HeH^+ + H \qquad \Delta H = 80 \text{ kJ mol}^{-1}$$

vibrational energy was found to be an order of magnitude more effective than translational in stimulating reaction, whereas electronic excitation of the ion is most efficient in promoting:

$$N_2^+ + N_2 \rightarrow N_3^+ + N$$

However, in all cases the relative effectiveness of translational energy increases as energy is raised.

With exothermic reactions, in the absence of a substantial energy barrier, the role of excess reactant energy must be very different. For translational energy, the cross section for the exothermic reaction is maximum at low energy—considerably greater than the hard-sphere cross section attributed to the long-range ion–molecule attraction. As translational energy increases the threshold of competing endothermic reactions will be reached. The yield from the exothermic reaction will decrease and the total reaction cross section also decreases as at higher speeds the range of effective interaction diminishes.

The role of internal energy in promoting exothermic reactions has not yet been clarified, and there are some confusing experimental results. Vibrational excitation of the ion has little effect in the reaction:

$$NH_3^+ + H_2O \rightarrow NH_4^+ + OH$$

The rate of the reaction

$$H_2^+ + H_2 \rightarrow H_3^+ + H$$

decreases with ion vibrational excitation but the reverse effect is found for vibrationally excited neutral molecules in:

$$O^+ + N_2 \rightarrow NO^+ + N$$

The reaction:

$$Ar^+ + H_2 \rightarrow ArH^+ + H$$

has a very high cross section but it is even greater for the electronically excited ion.

10.10 Chemistry of the Upper Atmosphere[20]

As altitude above the earth's surface increases atmospheric pressure decreases, but the variation of temperature is rather more complicated. It decreases initially but then increases, decreases again and finally rises steadily all the way to the sun. These

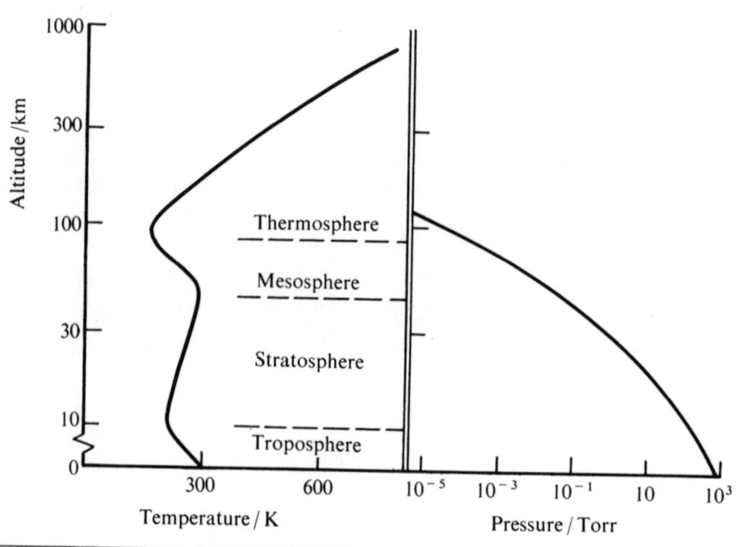

Figure 10.4
Variation of atmospheric temperature and pressure (on log scale) with altitude (on log scale), showing the various regions of the atmosphere.

[20] For useful reviews, see T. M. Donahue, *Science*, **159**, 489 (1968); M. J. McEwan and L. F. Phillips, *Accts Chem. Res.*, **3**, 9 (1970).

features, illustrated in figure 10.4, mean that there are several well-defined layers of the atmosphere and these are indicated on the diagram. The atmosphere also contains ions and electrons and the electrically charged region, particularly above 60 km altitude where concentration of charged species increases, is called the *ionosphere.*

The atmosphere is predominantly nitrogen and oxygen together with minor constituents such as carbon dioxide, the rare gases, and varying amounts of water vapour. At increasing altitudes, as the atmospheric pressure decreases the fraction of some important minor components increases until they dominate the chemistry of the system.

Which species predominate at different altitudes depends on a variety of factors, especially the absorption of the sun's radiation. At the higher altitudes low wavelength, extremely energetic quanta cause photoionization of molecules giving ions and electrons. At somewhat lower levels absorption of longer wavelength ultraviolet radiation gives atoms and radicals, the last of the ultraviolet being absorbed by ozone which leaves only visible and longer wavelength radiation to reach the surface. The system is further complicated because some of the photochemical processes and ionic and radical reactions can give species in excited energy states which can be quite long lived at low pressure.

Some aspects of atmospheric chemistry have already been mentioned—pollution of the troposphere in section 7.11 and reactions of ozone in the stratosphere in section 6.2. In the upper atmosphere ion–molecule reactions are of the greatest importance and will be discussed here.

Investigating the ionosphere

The ionosphere was discovered through the way the electrons reflected radio waves and such techniques are used to probe the behaviour of electrons up to the region of maximum electronic concentration. Above this layer satellite or rocket borne experiments are required and the analytic instruments, optical spectrometers and mass spectrometers, provide information on concentrations of ions and neutral species and details of the sun's u.v. spectrum. Figure 10.5 shows the variation with altitude of the concentration of ions and electrons. So many ions have been identified and their possible interactions are of such number that detailed laboratory investigations of ion–molecule reactions provide essential evidence to help interpret the system. There has been considerable success in identifying the main processes taking place but we are far from a comprehensive understanding of the upper atmosphere.

Another interesting phenomenon is the emission of light which can be seen from rockets in daytime and is visible from the ground at night (nightglow). This must originate in chemiluminescent processes.

Formation of ions and radicals

The main source of ions and electrons is photoionization, the absorption of X-rays and far u.v. radiation by oxygen to give O_2^+ and N_2^+. Some of the photoelectrons are liberated with sufficient energy to cause further ionization when they collide with other molecules.

Though nitrogen molecules do not undergo appreciable photochemical decomposition, oxygen molecules absorb longer wavelength u.v. light and dissociate to atoms. The

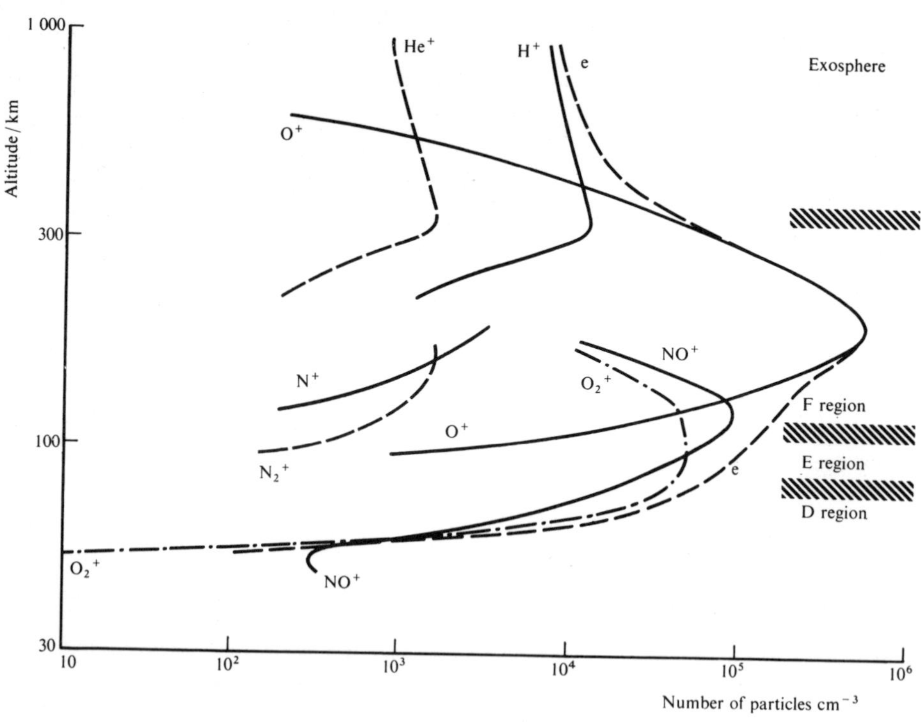

Figure 10.5
Variation in concentration of electrons and important ions with altitude in the ionosphere in daytime. Both axes are plotted on logarithmic scales.

relative concentration of oxygen atoms increases with altitude until at 120 km they are more abundant than molecular oxygen, and at 200 km more abundant than molecular nitrogen. At higher altitudes still hydrogen atoms, formed by photolysis of water molecules, and helium atoms become the major constituents. These atoms are also ionized at high altitudes.

If these are the primary species formed, other positive ions, excited ions, excited molecules and radicals are produced in various ion–molecule reactions. The ultimate fate of ions is neutralization by recombination with electrons. Less is known about negative ions which are more difficult to detect since the rockets which sample ionic concentrations themselves build up negative charge in flight thus repelling the negative ions.

Figure 10.5 shows the regions into which the ionosphere is conventionally divided —D, E, F and exosphere. The important processes in the D, E, F regions will be examined in the remainder of this section. For the exosphere it will suffice here to state that it is rather poorly understood and the concentrations of the major constituents He^+ and H^+ are difficult to explain on present knowledge of the rates of processes in which they are believed to be formed and removed.

F and upper E region

The main observations can be summarized: there is an increasing predominance of O^+ which reaches a maximum about 100 km above the maximum atom concentration; N_2^+ plays a minor role although it is formed in greater abundance than many other species; NO^+ is a major constituent even though the concentration of neutral nitric oxide is very small.

One factor determining the relative concentrations is their respective neutralization rates. This rate is quite high for N_2^+, O_2^+ and NO^+, but comparatively low for O^+. This maintains the high concentration of O^+, which is removed at lower altitudes mainly in the reactions:

$$O^+ + O_2 \rightarrow O_2^+ + O$$

$$O^+ + N_2 \rightarrow NO^+ + N$$

O^+ concentration falls at very high altitude because, at the very low total pressures, it is dispersed mainly by diffusion.

There are other important reactions which remove N_2^+ and form O_2^+ and NO^+:

$$N_2^+ + O_2 \rightarrow O_2^+ + N_2 \qquad (10.10.I)$$

$$N_2^+ + O \rightarrow NO^+ + N$$

These reactions account for the depletion of N_2^+ and the abundance of NO^+, which is further enhanced because it has a very low ionization energy and does not undergo charge transfer on collision.

D and lower E region

Here atmospheric pressure is significantly higher. N_2^+ and O_2^+ are again the primary ions formed by photoionization, and again there are reactions such as (10.10.I) diminishing N_2^+ and enhancing O_2^+ concentration. However, the latter concentration is much higher than expected given the reactions in which it could be removed, e.g.

$$O_2^+ + NO \rightarrow O_2 + NO^+$$

$$O_2^+ + N_2 \rightarrow NO^+ + NO$$

With the higher atmospheric pressure in the D region negative ions of O_2 could be formed in high concentration. Laboratory experiments suggest there should be important reactions involving these ions, e.g.

$$O_2^- + H_2O + M \rightarrow H_2O.O_2^- + M$$

$$O_2^- + CO_2 + M \rightarrow CO_4^- + M$$

However, due to the experimental difficulties in measuring negative ion concentrations it is not even certain what species exist there.

Minor atmospheric constituents must also be playing some part in ion formation. There must be processes involving hydrogen-containing molecules and radicals, ozone, carbon and nitrogen oxides and even ice or dust, and these are not yet understood. There is certainly a range of ion–molecule clusters, e.g. $H_3O^+ . H_2O$ and higher hydrates, detected by rocket borne mass spectrometers and their formation has not been explained satisfactorily.

Sometimes at altitudes 90–95 km there is a sharp rise in the concentration of ions such as Na^+, Si^+, Ca^+, Mg^+. The source appears to be meteoric dust which is subject to complicated air transport processes. This level is called the *sporadic E layer*.

Nightglow

Nightglow emission comes from chemiluminescent reactions such as

$$H + O_3 \rightarrow OH^{\dagger} + O_2$$

$$O + O + M \rightarrow O_2{}^{\dagger} + M$$

Reactions of the latter type give several different excited states of O_2 which emit in different regions of the spectrum. There is also weak emission from atomic hydrogen. The nightglow may explain the observation that during the night, in the absence of photoionization by the sun, the concentrations of $O_2{}^+$ and NO^+ do not fall as much as expected at altitudes near 130 km. Emission from atomic hydrogen at different wavelengths could be responsible for ionizing NO and O_2, and charge transfer between $O_2{}^+$ and NO could provide extra NO^+.

Further Reading

(A)

T. Carrington and D. Garvin, *Comprehensive Chemical Kinetics*, Vol. 3, (Eds C. H. Bamford and C. F. H. Tipper), Elsevier, 1969.

T. Carrington and J. C. Polanyi in *MTP Int. Rev. Sci., Phys. Chem., Ser. 1*, Vol. 9, (Ed. J. C. Polanyi), Butterworths, 1972.

J. C. Polanyi, *Accts Chem. Res.*, **5**, 161 (1972).

(B)

R. L. Wolfgang, *Prog. Reaction Kinetics*, **3**, 97 (1965).

F. S. Rowland in *MTP Int. Rev. Sci., Phys. Chem., Ser. 1*, Vol. 9, (Ed. J. C. Polanyi), Butterworths, 1972.

(C)

Ion–Molecule Reactions, (Ed. J. L. Franklin), Plenum, 1972.

B. H. Mahan, *Accts Chem. Res.*, **1**, 217 (1968); *Accts Chem. Res.*, **3**, 393 (1970).

T. M. Donahue, *Science*, **159**, 489 (1968).

Appendix

Basic units

Physical quantity	Name of unit	Symbol
length	metre	m
mass	kilogramme	kg
time	second	s
electric current	ampere	A
temperature	kelvin	K
amount of substance†	mole	mol

† Recommended additional unit.

Derived units

Physical quantity	Name of unit	Symbol	Definition
energy	joule	J	$kg\ m^2\ s^{-2}$
force	newton	N	$J\ m^{-1}$
electric charge	coulomb	C	A s
potential difference	volt	V	$J\ A^{-1}\ s^{-1}$
frequency	hertz	Hz	s^{-1}
pressure	pascal†	Pa†	$N\ m^{-2}$

† Not widely used.

Prefixes for SI units

Prefix	Symbol	Factor
mega	M	10^6
kilo	k	10^3
deci†	d	10^{-1}
centi†	c	10^{-2}
milli	m	10^{-3}
micro	μ	10^{-6}
nano	n	10^{-9}
pico	p	10^{-12}

† Should be used only where recommended prefixes are inconvenient; centimetre (cm) is particularly useful (in gas kinetics).

Conversion factors

Physical quantity	Unit	Symbol	SI equivalent
length	ångstrom	Å	10^{-10} m
volume	litre	1	10^{-3} m^3
	cubic cm	cm^3	10^{-6} m^3
energy	electron volt	eV	1·602 10 × 10^{-19} J
		erg	10^{-7} J
	calorie (thermochemical)	cal	4·184 J
pressure	atmosphere (standard)	atm	101 325 N m^{-2}
		Torr	133·322 N m^{-2}
		(1 atm = 760·00 Torr)	

Physical constants

Constant	Symbol	Value
gas constant	R	8·314 3 J K^{-1} mol^{-1}
Boltzmann constant	$\bar{k}$	1·380 54 × 10^{-23} J K^{-1}
Avogadro constant	L	6·022 52 × 10^{23} mol^{-1}
Planck constant	h	6·625 6 × 10^{-34} J s
velocity of light	c_0	2·997 925 × 10^8 m s^{-1}
mass of electron	m_e	9·109 1 × 10^{-31} kg
charge of electron	e	1·602 10 × 10^{-19} C
molar volume of ideal gas at 101 325 N m^{-2}, 273·15 K		22 413·6 cm^3 mol^{-1}

Index of Names

Index of Subjects